Lead Regulation Handbook

Edward E. Shea

Government Institutes
Lanham, Maryland

Government Institutes

Published in the United States of America
by Government Institutes, an imprint of The Scarecrow Press, Inc.
A wholly owned subsidiary of
The Rowman & Littlefield Publishing Group, Inc.
4501 Forbes Boulevard, Suite 200
Lanham, Maryland 20706
http://www.govinstpress.com/

Estover Road
Plymouth PL6 7PY
United Kingdom

British Library Cataloguing in Publication Information Available

Library of Congress Cataloging-in-Publication Data

Shea, Edward E., 1932–
Lead regulation handbook / Edward E. Shea. — 2nd ed.
p. cm.
Includes index.
ISBN-13: 978-0-86587-161-8 (pbk. : alk. paper)
ISBN-10: 0-86587-161-2 (pbk. : alk. paper)
1. Lead—Law and legislation—United States. I. Title.
KF3964.L43S53 2007
344.7304'633—dc22 2007007381

™ The paper used in this publication meets the minimum requirements of American National Standard for Information Sciences—Permanence of Paper for Printed Library Materials, ANSI/NISO Z39.48-1992.
Manufactured in the United States of America.

THE AUTHOR

Edward E. Shea is a partner and heads the environmental practice of Windels Marx Lane & Mittendorf, LLP, a law firm with offices in the City of New York and in New Brunswick, New Jersey.

Mr. Shea formerly served for over ten years as a senior executive, general counsel and a director of Reichhold Chemicals, Inc., a Fortune 500 multinational manufacturer of chemicals and related products, and in similar capacities for four years for GAF Corporation, a multinational manufacturer of specialty chemicals and building materials. Mr. Shea also served for eight years as an officer and counsel to the Peridot Chemicals Group in addition to his law practice.

Mr. Shea is the author of the book "Environmental Law and Compliance Methods," Oceana, 2002 and several articles in the National Law Journal, the New York Law Journal, the Practical Real Estate Lawyer, Environmental Compliance & Litigation Strategy and other publications. He is also an Adjunct Professor in the Graduate School of Business of Pace University in New York.

The author has been familiar with lead all of his life. As a child, the author and other children were given toy automobiles and soldiers made of lead and painted with lead-based paint. As a teenager, the author helped his father paint the family home with lead-based paint. In the U.S. Navy, the author chipped and applied lead-based paint to the bulkheads of a battleship. As a young lawyer, the author worked with financial printers who set hot lead type for prospectuses to sell stocks and bonds. In business, the author helped to establish compliance with environmental, safety and health regulations in plants producing resins and pigments for the paint industry. As an environmental practitioner, the author works with projects for the abatement of lead in soil and groundwater and of lead-based paint as well as compliance with health and safety regulations.

PREFACE

The Lead Regulation Handbook was the first book to present an overview of the numerous laws, regulations and court decisions applicable to regulation of lead and its compounds and alloys. The Second Edition updates the original book.

Chapters I to III acquaint the reader with the characteristics of lead and how it is mined, smelted, refined and sold for many different uses including some applications which protect human safety and health. These Chapters now include updated information on production of lead and lead-containing products including widespread growth of battery recycling and production of secondary lead smelting throughout the United States.

Chapter IV describes the adverse health effects that can result from excessive exposure to lead. Special focus is placed on the epidemiological and other studies and surveys of low-level exposure of very young children that are the subject of widespread debate. The Chapter now describes numerous new studies, domestic and international, of health effects of lead including cognitive effects on very young children and attention deficit hyperactivity disorder (ADHD); studies on remediation methods; studies on treatment by chelation; and statements by the Centers for Disease Control and the American Academy of Pediatrics.

Chapters V to XIX discuss regulation of lead and its compounds and alloys under each of the major environmental, safety and health laws. These chapters have been updated to include, among other things, (1) EPA's national emission standards for primary and secondary lead smelting and consideration of a new ambient air quality standard under the Clean Air Act; (2) EPA's water quality regulations under the Clean Water Act; (3) EPA's lead and copper regulations under the Safe Drinking Water Act; (4) EPA's revised reporting thresholds and classification of lead as a persistent, bioaccumulative or toxic chemical under the Emergency Planning and Community Right to Know Act; (5) EPA's amendment of its municipal landfill regulations under the Resource Conservation and Recovery Act to allow disposal of lead abatement waste in construction and demolition (C&D) landfills and its adoption of a universal waste standard and an exemption for lead-acid batteries being reclaimed; (6) EPA's adoption of worker protection and other regulations under the Comprehensive, Environmental Response, Compensation and Liability Act; (7) updated descriptions of OSHA's general industry and construction standards for lead and actions to enforce them; (8) updated information on CPSC's regulatory programs, including children's metal jewelry, under the Consumer Product Safety Act; (9) updated information on FDA's food and cosmetics additive regulations under the Federal Food, Drug, and Cosmetics Act including FDA's guidance actions on lead in children's candy and on folk remedies and cosmetics; (10) reorganization of the USDOT which administers the Hazardous Materials Transportation Act; (11) updated descriptions of HUD's numerous housing programs and their lead-based paint requirements; and (12) updated descriptions of methods for evaluation, inspection and control or abatement of lead-based paint including the HUD Guidelines.

Chapters XX to XXIII are devoted to state laws and regulations; lead-based paint abatement including Guidelines published by the U.S. Department of Housing and Urban Development; national non-government standards organizations; and the growth of litigation resulting from the deterioration of lead-based paint in residential dwellings in older urban areas of the nation. These Chapters have been updated to provide citations to numerous new state laws, updated information on ASTM International, Inc. and other leading standards organizations.

Chapter XXII provides discussions of important new developments in lead-based paint litigation including the major lawsuits based on a public nuisance theory brought by State attorneys general against former manufacturers of lead pigments and their trade association.

The chapters provide citations to statutes, regulations, court decisions, scientific studies and surveys, and industry publications that enable interested readers to pursue the topics further.

NOTE: Just as the final manuscript for this book was being prepared, researchers from the schools of public health at several major universities in the United States and health institutes in Mexico published a mini-monograph on adult lead exposure. The mini-monograph consists of five articles which are detailed literature reviews on subjects including the biomarkers used to measure lead dosage, medical management of adult lead exposure, cardiovascular disease, and cumulative lead dose and cognitive function in adults. The introductory article calls for modifications of OSHA's personnel exposure limits for lead to lower concentrations that will protect against both acute health effects of recent dose and chronic health effects of cumulative dose. The mini-monograph is discussed further on page 56 of Chapter IV. See "Adult Lead Exposure: Time for Change," Schwartz and Hu, Environmental Health Perspectives, Vol. 115, No. 3, p. 451, March 2007.

Edward E. Shea

THE LEAD REGULATION HANDBOOK

Table of Contents

APPENDICES:

I. LEAD - GENERAL BACKGROUND

A. History

Lead has been a useful material for at least 5,000 years. Archaeologists have identified early uses in Egypt and Asia Minor about 3,000 B.C. Lead was extensively used in Greece, Italy and Spain and was an important contributor to the economy in Athens, Rome and Constantinople. Early uses of lead were in building materials, water piping and pigments for glazing. The castles and cathedrals of Europe contain extensive uses of lead in roofs, windows, pipes and decorations. Lead was also widely used in the early years of the United States in building materials, water piping, pewter, ammunition and solder.

B. Properties of Lead

Lead has a valence of 2 or 4 and is a component of many chemical compounds. For example, reacting with a valence of 2, lead will form the compound lead oxide (PbO), also called litharge. Reacting with a valence of 4, lead will form the compound lead dioxide (PbO_2).

Lead is a soft heavy blue-gray metal. It is an element found in group IV-A of the periodic table. Its atomic number is 82. Some important physical characteristics are as follows:

Atomic weight	207.2
Melting point	327.4 °C (621 °F)
Boiling point	1770 °C
Density (cast lead)	11.33 g/cm^3 at 20 °C (68 °F)
Weight (cast lead)	0.41 lbs. per in^3
Volume	1 lb. = 2.44 in^3
Linear expansion coefficient	29.1 x 10^{-6} at 20 °C
Thermal conductivity	34.7 W/(nk) at 28 °C
Tensile strength, k pa	14,000 at 20°C
Hardness (Moh's scale)	1.5
Brinell, No.1 cm ball, 30 sec. 100 kg load	3.2 to 4.5
Electrical resistivity uΩ/cm	20.65 at 20°C
Specific conductance, d(Ωm)-1	4.83 x 10^4 at 18°C

Lead is very dense, soft, relatively ductile, malleable, corrosion resistant, and resistant to certain acids such as sulfuric acid, chromic acid, phosphoric acid, hydrochloric acid and hydrofluoric acid. However, lead dissolves in acetic acid and nitric acid and is attacked by caustic materials.

Lead (unalloyed) has little strength and is not elastic. Although its corrosion resistance is partially attributable to excellent water resistance, it is slightly soluble in water.

Lead has useful heat and electricity resistance and conduction characteristics. Lead acts as a barrier to sound and radiation.

C. Sources and Production of Lead

Lead is found in many areas of the world in naturally occurring ores, primarily as galena (lead sulfide), but also as cerussite (lead carbonate) and anglesite (lead sulfate). Among the leaders in mine production are Australia, Canada, China, the United States, Peru, Mexico, Sweden, the nations formerly comprising the Soviet Union and Yugoslavia. World annual mine production increased rapidly during the twentieth century, but has been at levels of about three to four million tons since the 1970s because a growing volume of lead has been produced by secondary recovery facilities.

In the United States, mine production of lead is now in southeastern Missouri. However, lead has also been mined in Alaska, Arizona, Colorado, Idaho, Montana, Nevada, New Mexico and Washington.

Ore bodies of commercial interest range from 2% to 6% lead and must undergo drossing, smelting, drossing and refining processes to produce lead for commercial use. In primary smelting operations, the impurities removed include sulfur and a variety of metals such as zinc, silver, copper, antimony, bismuth and arsenic. In recent years, scrap materials containing lead or lead alloys have provided about half of the raw materials used in lead production throughout the world and a great majority of the raw materials used in the United States. They are reclaimed in secondary smelting operations.

D. Sales and Consumption of Lead

World production and consumption of refined lead probably exceeds five million tons per year. Refined lead sold for commercial use must meet specifications calling for purity of at least 99.85%. Lead is regularly produced for commercial sale at levels exceeding 99.99% purity.

E. Alloys, Compounds and Isotopes

Lead is frequently mixed with other metals to produce alloys. Lead is reacted with other organic and inorganic substances to form compounds. See Chapter III.

Lead is also found in nature as isotopes. One is lead 204. Others are lead 206, 207 and 208 which are radioactive. Lead 206 and 207 result from the decay of uranium. Lead 208 results from the decay of thorium. Other lead isotopes, which are unstable radioactive materials, are lead 200 to 203, 205, and 209 to 214. See Chapter XVIII.

NOTE: Information in this Chapter was based on various sources including the Kirk-Othmer Encyclopedia of Chemical Technology, Fifth Edition (2005); Ullman's Encyclopedia of Industrial Chemistry, Sixth Edition (2002); and publications of The Lead Industry Association, Inc.

II. MODERN USES OF LEAD AND LEAD COMPOUNDS

A. Overview

Because of its physical characteristics and its ability to form numerous alloys and compounds, lead is used in many commercial products:

Lead-Acid Batteries: During the past century, lead-acid batteries have been the most important use of lead, accounting in recent years for well over half of the annual tonnage produced in the United States. In a basic battery, lead alloy plates serve as the negative and positive terminals with sulfuric acid as the electrolyte. Lead-acid batteries are important as a clean quiet source of electric power for industrial transportation. Large storage batteries are also being used as a backup supply when electrical power outages occur. Batteries are discussed in more detail later in this Chapter.

Radiation Shielding: Lead is used in a variety of radiation protection applications where its density serves as a shield. For example, lead is used with X-ray medical imaging and radiation equipment. Leaded poly(vinyl chloride) plastic, commonly called "vinyl," material provides both radiation protection and sound proofing in aircraft. Lead is used as shielding in nuclear power plants and in containers for storage and disposal of radioactive waste. Radiation shielding is discussed in more detail later in this Chapter.

Electronics: Lead alloy solders are used in the manufacture of electronic components. Lead glazes are used to encapsulate electronic microcircuits, protecting them from corrosion. A lead compound is being used in place of silicon in advanced computer chips. Use of lead in electronic components is discussed in more detail later in this Chapter.

Glass: Lead is used in the glass picture tubes of all television sets and personal computer monitors in order to provide protection from X-rays. Lead is also a component of fine crystal glassware. Use of lead in glass is further discussed later in this Chapter.

Acid Production and Electrostatic Precipitators: Because of its resistance to some acids, lead is used in the production, storage and transportation of sulfuric acid, chromic acid and phosphoric acid. Sulfuric acid and phosphoric acid are used in the manufacture of fertilizer products important to food production. Lead is used in electrostatic precipitators which remove sulfuric acid mist and sulfur dioxide from the vent stacks of electric power plants, thus helping to control conditions that may result in "acid rain". Electrostatic precipitators are discussed in more detail later in this Chapter.

Construction: Lead is used in construction for its waterproofing, soundproofing and vibration damping characteristics. For example, heavy machinery can be installed on lead pads to limit the effects of vibration. Lead has been used for centuries as a roofing material because of its resistance to water and corrosion. Cathedrals such as St. Paul's Cathedral in London and Sancta Sophia in Istanbul have lead roofs that are centuries old. While roofs made of asphalt, glass fiber and mineral rock are most widely used in the United States, roofing systems containing lead are used where high quality and long life are important. They will usually be used with overlay and underlay materials containing asphalt, bitumen or rubber. Uses of lead in construction are described in more detail later in this Chapter.

Telephone and Television Cables: Lead alloys are used as sheathing for telephone and television cables because its ductility allows extrusion around the cables. This use has been limited by competition from plastic and other materials, but continues, especially where corrosion resistance is important. For example, lead was used as sheathing for electrical cable in the Channel Tunnel between England and France.

Coatings: Lead compounds have been used for many years as pigments and driers in coatings. Lead chromate ($PbCrO_4$), commonly called chrome yellow, makes high visibility highway marking paint to enhance traffic safety because it has good service life and effectively covers dark underlying materials. Molybdate orange ($PbMoO_4 \cdot PbCrO_4 \cdot PbSO_4$) is used as primer for bridges and other steel structures. As discussed later, sale of lead-containing paint for consumer use was banned in 1978, but use for industrial purposes continues to be allowed. For example, coatings containing lead pigments protect bridges and other steel structures from rust and corrosion caused by weather and road salt significantly longer than coatings containing other color additives. A variety of lead compounds (such as lead acetate, lead naphthenate, lead octoate and lead stearate) have been used as drying agents in solvent-based paints, but their use was reduced by the ban on consumer sale of lead-based paints and the growing use of water-based emulsion paints. Problems resulting from the continued presence of lead-based paint in millions of older residences are discussed later in this book. Titanium dioxide replaced lead in white pigment decades ago.

Ammunition and Detonators: Lead continues to be used in significant quantities as ammunition. Lead azide, $Pb(N_2)_3$, is an explosives detonator.

Tetraethyl Lead: Lead is a component of tetraethyl lead, a fuel additive used to reduce engine knock. The Clean Air Act established programs which gradually introduced unleaded fuels and phased out fuels containing tetraethyl lead from use in the United States at the end of 1995.

Water Piping and Solder: Lead was formerly used in water piping and solder, but has been restricted for these uses by federal and state laws for many years. Problems resulting from the continued existence of old lead water piping and solder are discussed later in this book. Lead continues to be used in plumbing applications where drinking water is not affected, usually with a coating or other underlay and overlay material.

Other Uses: Lead has been used to make cames to hold stained glass for centuries. Lead is also used in ceramic materials and glazes. Lead is used in fishing sinkers. Lead alloys were used as printer's type, but modern typesetting uses different technologies.

B. Lead-Acid Batteries

Lead-acid batteries account for over 60% of the lead used in the United States and the world. They are an important part of the present economy and plans for a future industrial economy that will be compatible with protection of human health and the environment.

Batteries can be grouped in several categories. One classification focuses on their use to provide:

(A) starting, lighting and ignition (SLI) for motor vehicles powered by internal combustion engines;

(B) traction power for short range electric vehicles such as fork lift trucks; and

(C) standby industrial power sources for lighting, electronical, computer, process control, air traffic control, hospitals and other systems.

A typical industrial lead-acid battery is contained in a polypropylene case and consists of several cells, each generating two volts. Large batteries can generate up to 72 volts.

The battery case contains poles, grids, separators and sulfuric acid electrolyte. The poles and grids (or plates) are made of lead alloy. The positive plates are coated with a lead peroxide paste. The sulfuric acid reacts with the lead peroxide in the positive plates and the elemental lead in the negative plates to form lead sulfate on both plates. Oxygen from the lead peroxide on the positive plates combines with hydrogen from the sulfuric acid to form water, producing electricity in the process.

The continuing reaction between the electrolyte and the plates causes the plates to become similar and the electrolyte to be diluted with water. The battery's capacity to generate electricity is reduced and eventually the battery is discharged. To prevent full discharge, batteries are recharged (usually at an 80% discharge level) by attaching them to an electric power source. The introduction of electricity reverses the reactions described above and restores the battery's capability to generate electricity.

Batteries have several advantages over internal combustion (IC) engines. They operate quietly and cleanly. They are relatively simple to operate and maintain. They are relatively economical and, with proper maintenance, have long service lives. They have environmental benefits because they do not generate dust, fumes, odors or significant heat. They do not burn fossil fuels.

Batteries also have several disadvantages. They are heavy, primarily because of their lead components. They require recharging at more frequent intervals and for longer time periods than is required for internal combustion (IC) engines. Vehicles powered with batteries must be operated at lower overall speeds. For these reasons, even with numerous improvements, batteries have not succeeded in matching the performance of IC engines for passenger vehicles or over-the-road industrial vehicles, ships or aircraft.

On the other hand, batteries are widely used in industrial vehicles operating short distances within industrial sites. They are specially useful inside factories and warehouses where their quiet, emission free operation is a significant advantage. For lift trucks, the heavy battery weight becomes an advantage in counterbalancing loads being transported and in providing traction on ice or snow. They can be operated safely on smaller areas. Their lower cost permits a factory to maintain a charging station where batteries can be charged and recharged.

Federal and state regulators, including the California Air Resources Board (CARB) continue programs designed to require automotive manufacturers to reduce vapor and exhaust emissions. One alternative is to develop commercially salable battery-powered electric automobiles for popular use. Thus far, however, battery weight and battery recharge frequency have been deterrents that led automotive manufacturers to turn to other alternatives such as hybrid automobiles.

C. Radiation Shielding

Radiation shielding is one of the most valuable uses of lead. During the past century, scientific and industrial research developed a wide range of useful radioactive equipment and materials. Examples are the X-ray machines, including CAT scan equipment, used for medical examinations. Another example is the radiation treatments used to retard the progress of cancer. Specially designed X-ray equipment is used to control and measure production in manufacturing plants. Nuclear fuel is used to generate steam in many electrical generating stations. The safe use of radioactive equipment and materials requires an effective shield from ionizing radiation to protect human health and the environment.

Radiation includes (1) electromagnetic waves consisting of X-rays and gamma rays, and (2) particle emissions of alpha particles and beta particles from radioactive materials or neutrons from a nuclear reactor. Shielding from alpha and beta particles is not difficult because common substances, such as wood, absorb them. However, X-rays, gamma rays and neutrons penetrate ordinary materials and pass through them. Thus, only materials which can attenuate their progress and also meet other requirements can be used as shielding for X-rays, gamma rays and neutrons.

Lead has several important shielding characteristics. X-rays and gamma rays are attenuated by dense materials. Lead is the densest of any material commonly available at low cost. Lead shielding attenuates X-rays and gamma rays. Lead shielding also attenuates neutrons and, unlike some materials, does not absorb neutrons in a manner that creates secondary radiation. Lead's conductivity enables it to carry away heat. Lead also does not suffer significant damage upon exposure to radiation, so it endures for long time periods and can be reused. Aggregate materials, such as concrete, can be produced to have density comparable to lead, but do not have its other important qualities. Thus, lead is the material of choice for radiation shielding.

Lead sheet, plate and foil are produced in various thicknesses for use in radiation shielding. However, since lead has little structural strength, it is also combined with other materials to produce lead laminates, lead plastic composites, lead bricks and blocks, lead glass and lead-filled acrylic sheet. Architects and building materials manufacturers have designed a variety of such products for use in the construction of hospitals and other medical or research facilities where radioactive equipment is used.

Lead is used with concrete to shield nuclear reactors in electrical generating stations. Lead is also used to shield nuclear reactors aboard submarines. Lead is also used in the casks designed for the storage and disposal of radioactive materials including nuclear fuel and low-level radioactive waste which requires transportation for disposal.

Several forms of lead used for radiation shielding are as follows:

FORM	USE
Lead Sheet, Slab and Plate	Permanent shield installations
Lead Shot	Where solid lead is impractical, due to location, shape and accessibility
Lead Wool	Filling deep cracks in a radiation barrier
Lead Epoxy	In-the-field crack filling patching
Lead Brick	Convenient, easily-handled; may be handled; may be moved and re-used
Lead Pipe	Shielding of radioactive liquids
Lead-clad Tubing	Shielding of radioactive liquids
Lead-lined/Lead-clad Pipe	Shielding of radioactive liquids
Lead Sleeves	Shielding of duct pipes carrying radioactive materials
Lead Powder	Dispersed in rubber or plastic for flexible shielding; also mixed with concrete and asbestos cement
Lead Glass	Transparent shielding
Lead-Polyethylene-Boron	Combined gamma, neutron and thermal neutron barrier material

D. Electronic Products

An important use of lead is in solders to join metal to metal, metal to glass, and metal to plastic in electronic circuitry.

The solders are alloys containing two or more metals, such as tin, in a percentage content ratio such as 63% tin and 27% lead. The alloys sold commercially include lead, tin, silver, copper, cadmium, aluminum, busimuth, iron, zinc and nickel. The metals and their content are adjusted to obtain desired characteristics such as melting temperature. The solders are used together with rosin, water soluble organic and solvent soluble organic fluxes. The fluxes are used for surface preparation, such as removal of oxides, so that the solder obtains a full adhesive bond.

Automated soldering equipment is widely used to permit many components to be soldered simultaneously to a printed circuit board. This is one capability which provides soldering with an advantage over other joining methods such as welding, brazing and mechanical fastening.

Recently, the electronics industry has developed computer chips which substitute lead zirconate titanate for silicon. The chips have the ability to retain data when power is turned off. Because these chips can resist cosmic radiation, the industry foresees their use in space and military applications.

E. Lead in Glass

Lead oxide performs a valuable protective function when used in glass for picture tubes in television sets and personal computer terminals because it provides shielding from X-rays without interfering with the transparency of the screen. Lead is used at high levels in glass in laboratory and control rooms in nuclear power and research facilities. Lead has also been used in glass for thermometers. Lead increases the refractive index of crystal glass, providing the sparkling appearance that enhances its use in crystal glassware and artistic works.

F. Lead in Electrostatic Precipitators

Electrostatic precipitators are widely used to protect the environment from emissions of sulfuric acid mist, fly ash and other air contaminants by electric power stations, pyrometallurgical smelters, petrochemical refineries, fertilizer plants and other heavy industrial plants. Public concern about "acid rain" has contributed to their use.

Because of its excellent resistance to sulfuric, chromic acid and phosphoric acid, lead and lead alloys are used for the upper and lower baskets, the tubes and as a covering for the steel core in electrostatic precipitators.

G. Architectural and Industrial Construction

As it has for centuries, lead continues to be used as a roofing material because of its excellent water resistance. In built-up roofing, lead sheet is used as a membrane, seams and flashing. The membrane is overlaid and underlaid with other materials such as asphalt.

Lead has also been widely used as waterproofing for construction of fountains and pools. In this application, lead sheet serves as the membrane and is overlaid and underlaid with other materials such as bitumen. Famous examples of pools using lead waterproofing are those at the World Trade Center, the United Nations Library and the American Airlines Building at JFK Airport in the City of New York, Cobo Hall in Detroit, and the Blue Cross Building in St. Louis.

Leaded vinyl sound barrier materials are used to protect workers from high noise levels created by industrial equipment. Grinding, cutting, crushing, pumping, high speed and other equipment often create noise that should be reduced to protect worker hearing and to comply with governmental noise control regulations. Since lead absorbs sound, noise can sometimes be reduced by laminating a thin layer of sheet lead to equipment, such as a grinder, which creates noise internally. In other situations, laminates of lead and vinyl sheet or vinyl foam can be used to enclose equipment or a production area. For example, curtains made of leaded vinyl can be hung around the equipment with a window through which the operator can monitor and control its operation.

NOTE: Information in this Chapter was based on various sources including publications of The Lead Industries Association, Inc., the Kirk-Othmer Encyclopedia of Chemical Technology, Fifth Edition (2005), Ullman's Encyclopedia of Industrial Chemistry, Sixth Edition (2002), and the author's experience working with industrial chemicals, coatings and pigments.

III. LEAD MINING AND PRODUCTION; ALLOYS AND COMPOUNDS

A. Primary Lead Production

1. Mining

Like other mining operations, lead mining advanced during the later years of the last century and the present century. For thousands of years, mining was performed in relatively shallow excavations by labor intensive operations. Mining is now generally performed at deeper levels by automated operations in which skilled workers operate highly productive equipment and controls. The lead mining industry developed some of the important advances that contributed to mining efficiency.

Deep mines are developed by the room and pillar method. A shaft provides access to the active mining level or levels. At the face or breast, the rock is broken by blasting with stable dynamite or ammonium nitrate/fuel oil (ANFO) products and by large drilling equipment. The ore is transported in front-end loaders, sometimes for considerable distances in large mines, to be dumped into skip pockets from which it is hoisted to the surface. At some mines, ore may be crushed before it is brought to the surface.

Among the advances of mining technology developed by the lead mining industry were the diamond drill and the use of roof bolts to secure the roofs of the rooms as ore recovery advances.

The advances in technology have included safety and protection of the environment. In previous centuries, mining was very dangerous work for many reasons. The rock breaking was done by humans with picks and shovels or primitive explosives. Hauling was done by mules, horses and humans. Hazards included collapses of rock and earth, falls, flooding, equipment injuries and the chronic effects of exposure to dust, gas and fumes. Modern, automated, safety-oriented mining methods have greatly reduced these hazards. In previous centuries and even during much of the present century, there were few requirements to reclaim land altered by mining activities. However, during recent decades, mining companies have reclaimed mined land to restore stability and normal vegetative growth and to reestablish conditions acceptable for human, animal and plant life.

Mine safety is regulated in the United States by the Mine Safety and Health Administration (MSHA), U.S. Department of Labor. MSHA's regulations are at 30 CFR Parts 1 to 104. Mining reclamation is regulated by the Office of Surface Mining Reclamation and Enforcement, U.S. Department of the Interior under programs which apply to surface coal mining, but also extend partially to non-coal reclamation. 30 CFR Parts 700 to 955. The states also have laws requiring reclamation of mine properties.

2. Concentration

Lead ore is either crushed to dimensions less than an inch in diameter underground or as the first step in the surface mill operations. The crushed ore is then ground in rod and ball mills to a small mesh size. The grinding process separates the metal bearing minerals from gangue. Flotation reagents may be added during grinding.

Before concentration, lead ores may contain 2% to 20% lead, but more commonly between 2% to 6% lead.

The ground ore slurry passes from the grinders to flotation equipment and frothing chemicals are added. The flotation equipment produces lead concentrate by separating lead ore from copper, zinc and silver ores and waste rock tailings. Some flotation processes are highly automated and controlled by X-ray analyzers that determine the metal contents of the feed, concentrates and tailings. At the end of the flotation process, the metal concentrates are filtered and dewatered. The tailings go to a setting pond and water is recirculated. The metal concentrates are now ready for sintering and smelting either by the same company for shipment to another company engaged in smelting one or more of the metal concentrates. The lead concentrate now contains about 75% lead.

3. Smelting

The lead concentrate which leaves the flotation process is still essentially galena, or lead sulfide. In the prevailing process, the next step is to remove the sulfur by sintering. The lead concentrate is mixed with flux materials and internally generated materials such as baghouse particulates. The mixture is pelletized and fed to a sinter machine to be roasted. A bed of the pellets is spread in layers on grates to make a sinter bed heated by burners. An updraft of air is passed through the sinter bed resulting in combustion production of lead oxide and sulfur dioxide.

The lead oxide is a hard porous "sinter cake" containing about 45% to 50% lead. The sinter cake is removed from the sinter machine and crushed to a size which can be fed to the blast furnace for reduction. The sulfur dioxide is commonly used to produce sulfuric acid. The sinter machine also produces dry and wet dust gases that must be controlled within air emission permit limits by baghouse or other collection equipment.

The crushed sinter cake is mixed with metallurgical grade coke and fed to a blast furnace. In the furnace, hot blasts of air up to 3,000 °F react with the carbon in the coke to form gases which reduce the sinter cake to molten lead bullion which flows into pots. The bullion still contains some other elemental metals and must be further refined. The blast furnace also produces slag which can be granulated and recycled or handled for disposal. Dusts and gases must either be further processed or controlled within the limits of air emission permits.

Some direct smelting reduction processes have been developed to overcome some of the disadvantages of the two-step sinter oxidation and blast furnace reduction processes such as high production of byproduct gases, high maintenance costs and the coke fuel used in the blast furnace. Some of these processes use dry charge flash smelting in an electric furnace.

Some well known smelting processes are the Imperial, Kivcot, QSL, Isasmelt, Kaldo (Boliden) and Outukumpu processes.

4. Drossing

After settling, concentrated bullion is transported in lead pots to a drossing process which removes copper by skimming and reskimming the dross containing the copper from the molten lead. A continuous process for copper drossing was developed by Broken Hill Associated Smelters Pty. Ltd. of Port Pirie, Australia.

5. Refining

Lead bullion from the smelting blast furnace and drossing process has a lead content ranging as high as 99%, but still contains a variety of metals and other impurities. They may include copper remaining after the drossing process, arsenic, tin, antimony, silver and other noble metals, zinc and bismuth, as well as other materials in trace quantities.

The refining processes are pyrometallurgical and include the following steps:

Copper Removal: Copper removal is a two-step process. In the first step, lead bullion is charged to a dross (reverberatory) furnace and copper matte is removed and may be sent to a copper smelter. In the second step, sulfur is introduced to form copper sulfide and the copper sulfide is removed.

Arsenic, Tin and Antimony Removal: Arsenic, tin and antimony can be removed by the Harris process which uses molten sodium hydroxide to oxidize them. Removal (also called softening) can also be accomplished by air injection into a furnace which contains molten lead bath. The air reacts with the tin, arsenic and antimony to produce floating slag which is removed from the furnace.

Noble Metals Removal: Silver and other noble metals can be removed by the Parkes process which involves precipitation with zinc. Silver may be recovered in quantities having commercial value.

Zinc Removal: Zinc is removed by vacuum distillation.

Bismuth Removal: Bismuth can be removed by the Kroll-Betterton process by precipitation with alkali metals or alkaline earth metals. If so, the alkali metals or alkaline earth metals will be removed by selective oxidation.

The purified lead (99.99%) is cast into blocks, ingots and smaller pigs and calking links for sale to customers or is further processed to make alloys and compounds.

Process flow sheets showing the mining, concentration, sintering, smelting, refining and casting of lead appears at the end of this Chapter.

B. Secondary Lead Production

1. Recycled Scrap Feed Materials

In recent years, about half of the total annual production of lead in the United States and elsewhere in the world was produced in secondary recovery facilities. These facilities process scrap materials which consist primarily of batteries which must be recycled in compliance with environmental laws and regulations. They may also process other lead scrap materials such as pipe and sheet.

2. Battery Breaking

The first challenge in secondary lead production is battery breaking in an efficient and environmentally acceptable manner. Batteries typically consist of a polypropylene plastic case containing lead and plastic components, including ebonite and separators, and sulfuric acid electrolyte. The lead components must be separated and recovered. The plastic components and acid must also be separated and either converted into useful byproducts or handled properly for disposal.

Batteries can be drained of acid and the whole battery charged to a blast furnace. However, plastic materials, which tend to fuse, can create flow problems in the furnace and produce volatile organic gases which require environmental control. Batteries can also be processed by removal of the case and the acid and using the contents as feedstock, but these steps lack efficiency. Thus, the first step is usually to feed the batteries to a hammer mill where they are mechanically shredded. The components are then separated by screening and gravity separation in water.

3. Smelting

After separation, posts and grids can be sent directly to a smelter to produce lead-antimony alloys. Battery parts can be desulfurized and processed in a reverberatory furnace to produce lead and high antimony slag which then goes to a blast furnace. Lead from the smelter and the two furnaces is then cast into ingots and blocks. If desired, antimony can be removed from the lead by oxidation as described earlier.

4. Byproducts

The propylene from the battery casings can be recycled or handled for disposal. The sulfuric acid can be reacted with sodium carbonate and crystallized to produce sodium sulfate for sale to manufacturers of detergents, paper or glass.

Exhaust gases from the smelter, reverberatory furnace, and blast furnace must be controlled by baghouse filtration equipment. Slag which cannot be further processed will be tested for toxic characteristic.

The process flow sheet showing the secondary production of lead by recycling of batteries and other scrap and a list of the kinds of scrap recycled by secondary producers in several areas of the nation appears at the end of this Chapter.

5. Battery Breaking Processes

Known process technologies for battery breaking and recovery of valuable materials were developed by Engitec of Milan, Italy and M.A. Industries of Atlanta, Georgia.

C. Lead Alloys

Lead has many useful properties. However, its tensile strength and creep strength are low. The strength of lead can be improved and other useful characteristics gained by combining lead with other metals to form alloys. Some of the common lead alloys are as follows:

- Lead-Antimony Alloys. Lead-antimony alloys have their major use as grids, posts and connectors in traditional batteries. A disadvantage is the tendency of antimony to break down water into hydrogen and oxygen during charging, thus discouraging use in sealed low maintenance batteries. Lead-antimony alloys are also used in cable sheathing, shot, wrought lead for acid storage, and other applications.
- Lead-Calcium Alloys. Lead-calcium alloys are used in modern automotive batteries and for standby power, submarine and sealed batteries.
- Lead-Tin Alloys. Lead-tin alloys are used primarily in solders to join and seal metals. Their use in electronic circuit boards is important and active. Their use in sealing steel (tin) cans has declined because of laws imposing prohibitions or restrictions and governmental programs to reduce lead content in food.
- Lead-Copper Alloys. Lead-copper alloys are used in cable sheathing. They are also used in lead sheets, pipe, wire, tank linings, tubes for acid mist precipitators, steam pipes, and roof sheathing.

- Lead-Silver Alloys. Lead-silver alloys are used in battery anodes and high temperature solders.
- Lead-Tellurium Alloys. Lead-tellurium alloys have high mechanical strength and are used in pipe, sheet, shielding for small nuclear reactors such as those aboard submarines, and cable sheathing.
- Reactive Lead Alloys. Lead-strontium alloys have been used in batteries, but their cost is high in comparison to lead-calcium alloys which have similar properties. Lead-lithium alloys have been used in bearings.
- Low-Temperature Alloys. Lead can be combined with metals having low melting points (such as tin, iridium, bismuth, silver or cadmium) to produce alloys with low melting temperatures for use in fuses, sprinkler systems and other applications where a low melting durable metal is needed. For example, lead-iridium alloys are used to make glass to metals seals and, because of their low vapor pressures, are used as seals in vacuum systems.
- High-Temperature Alloys. Lead can also be combined with other metals into alloys to have useful characteristics in relatively high temperature applications. One such composition is a tin/lead/silver eutectic (1 Sn/97.5 Pb/1.5 Ag).

D. Lead Compounds

Lead forms a wide variety of compounds. As described earlier, lead has two valences: 2 and 4. When formed, lead compounds may have more than one crystal structure, thus giving them different properties.

As previously described, lead is commonly found in nature as the compound lead sulfide which must be processed to obtain elemental lead. Elemental lead itself will not remain stable in the ambient atmosphere, but will form a surface layer of lead oxide followed by a film of lead carbonate that provides good water resistance.

Examples of lead compounds are as follows:

Lead Acetate. Lead acetates are formed by reacting lead with acetic acid. They are anhydrous lead acetate, lead subacetate, lead acetate trihydrate, and lead tetracetate.

Lead Carbonates. Lead carbonates are formed by reacting lead with carbon dioxide. They are lead carbonate and basic lead carbonate (white lead). Although the former widespread use of white lead as a pigment in paints has been discontinued, the lead carbonates have many other uses in plastics, glazes, inks and greases.

Lead Chromates. Lead chromates are compounds of lead, chromium and oxygen. They are used as pigments in industrial paints such as yellow traffic road marking paint and bridge primers.

Lead Halides. The lead halides are lead fluoride, lead chloride, lead bromide and lead iodide. They can be used in a wide variety of applications such as paint, batteries, glass, and highly specialized products. Lead tetrafluoride can be used to produce difluorocarbons and biologically active steroids.

Lead Oxides. The lead oxides are lead oxide, lead dioxide, lead trioxide (lead sesquioxide) and lead tetroxide (red lead). They are used in batteries, ceramics, rubber, plastics and other uses.

Lead Silicates. The lead silicates are lead monosilicate and lead bisilicate, which are used in ceramic glazes and glass, and tribasic lead silicate which is used in glass and frit.

Lead Sulfates. There are several lead sulfate compounds including tribasic lead sulfate and tetrabasic lead sulfate. Tribasic lead sulfate is used in the plastics industry as an additive to poly(vinyl chloride) compounds to provide heat stability.

Organolead Compounds. Lead forms a number of organolead compounds. By far the best known is tetraethyl lead which was used for many years as an antiknock gasoline additive, but has been prohibited in the United States by the Clean Air

Act. Lead also reacts with other alkyl and aryl groups to form a variety of components which have a number of commercial uses.

Lead Telluride. Lead telluride is made by melting lead and tellurium together and has semiconductive and photoconductive properties. It is used in pyrometry, heat sensing instruments, thermoelectric conversion elements, fuel cell catalysts, lithium battery cathodes, vacuum switch contacts, selective electrodes, tunable lasers, and thermistors.

NOTE: Information in this Chapter was based on various sources including publications of Doe Run Company and The Lead Industries Association, the Kirk-Othmer Encyclopedia of Chemical Technology, Fifth Edition (2005) and Ullman's Encyclopedia of Industrial Chemistry, Sixth Edition (2002).

The Mines:
Where Doe Run products begin ....

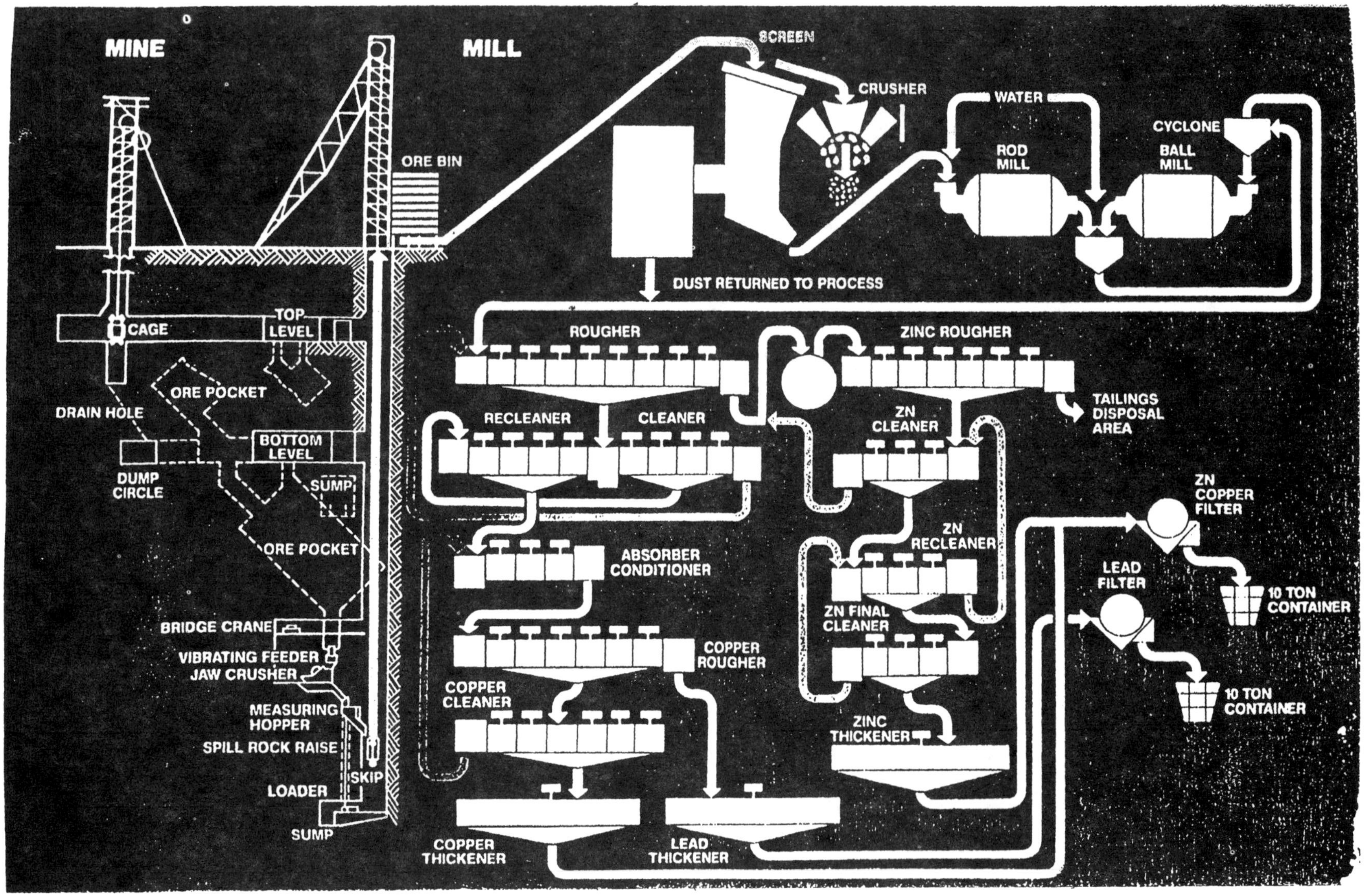

PROCESS FLOW SHEETS - PRIMARY LEAD PRODUCTION

..... continuation.....

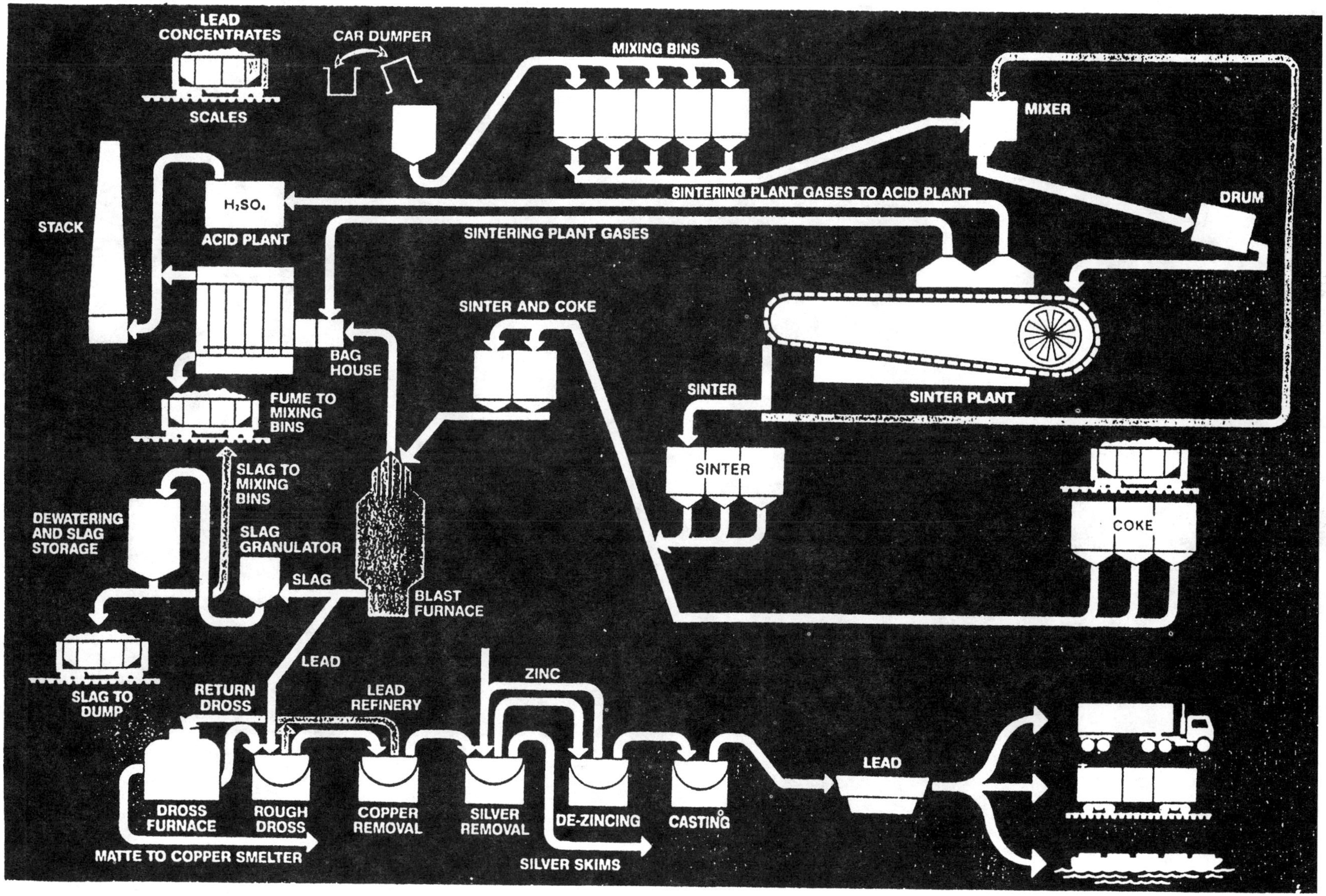

..... and magnification of process steps.

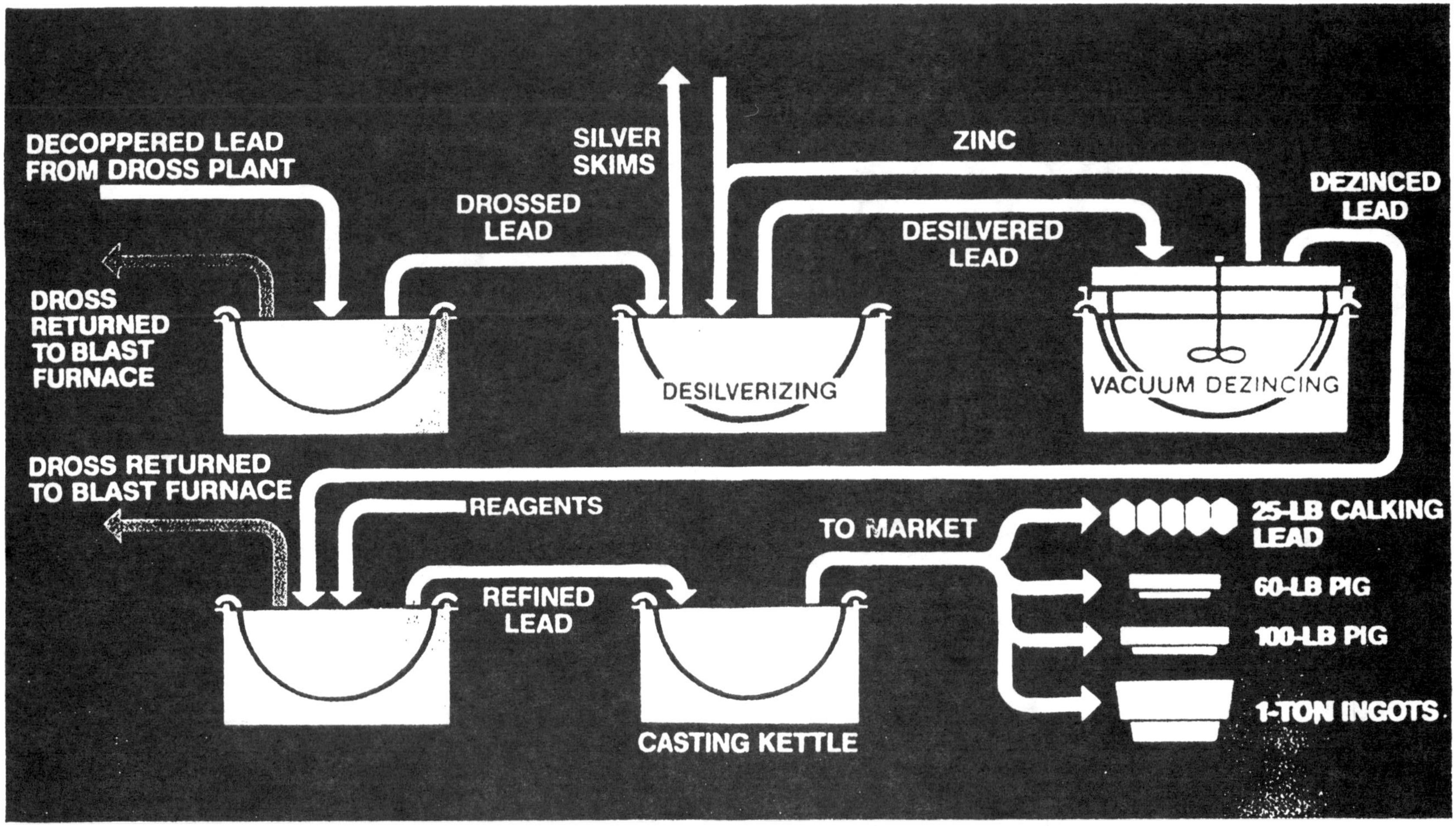

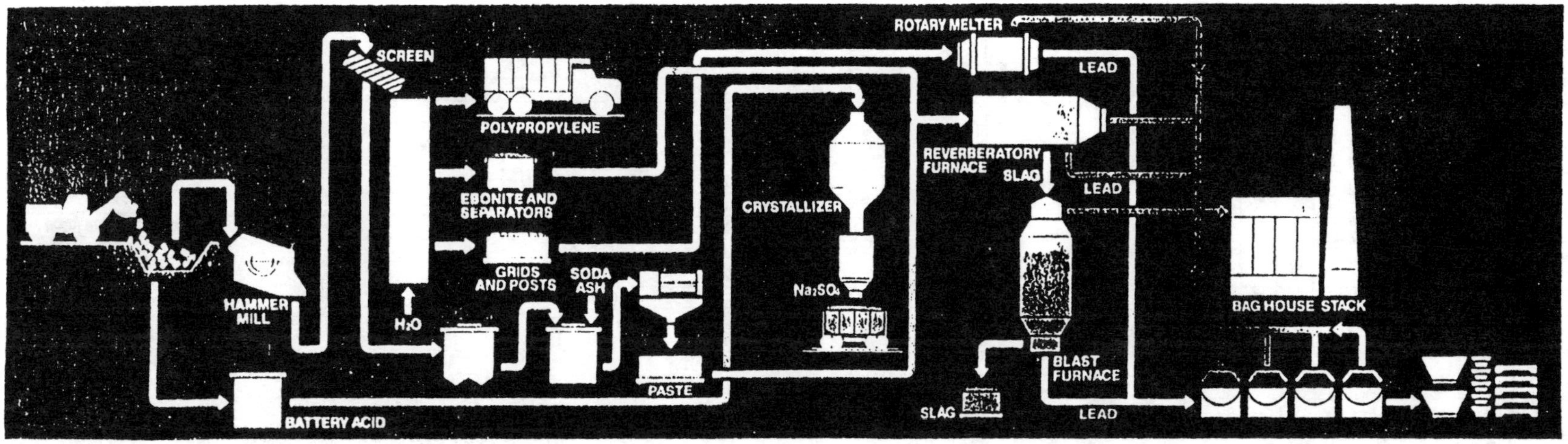

PROCESS FLOW SHEETS - SECONDARY LEAD PRODUCTION

FORMS OF LEAD SCRAP

Forms of Lead Scrap	Ames Metal Products Co. - Chicago, IL	ASARCO, Inc. - Glover, MO	ASARCO, Inc. - E. Helena, MT	ASARCO, Inc. - Omaha, NE	The G. A. Avril Co. - Cincinnati, OH	Canada Metal Co., Ltd. - Toronto, Canada	Canada Metal Co., Ltd. - Winnipeg, Manitoba	Cominco, Ltd. - Trail, British Columbia, Canada	The Doe Run Co. - Boss, MO	The Doe Run Co. - Herculaneum, MO	East Penn Mfg. Co., Inc. - Lyon Station, PA	Encycle Texas, Inc. - Corpus Christi, TX	Essex Metal Alloy Co. - Union, NJ	Exide Corp. - Muncie, IN	Exide Corp. - Reading, PA	Federated-Fry Metals - Altoona, PA	GNB Inc. - Columbus, GA	GNB Inc. - Frisco, TX	GNB Inc. - Vernon, CA	Gopher Smelting & Refining - Eagan, MN	Kinsbursky Brothers, Inc. - Anaheim, CA	Mayfield Manufacturing - Birmingham, AL	Metallic Resources, Inc. - Twinsburg, OH	Noranda Minerals - Belledune, New Brunswick, Canada	Nova Lead Inc. - Quebec, Canada	Refined Metals Corp. - Beech Grove, IN	Refined Metals Corp. - Memphis, TN	RSR Corporation - Middletown, NY	RSR Corporation - Indianapolis, IN	RSR Corporation - City of Industry, CA	Sanders Lead Co., Inc. - Troy, AL	Schuylkill Metals Corporation - Baton Rouge, LA	Schuylkill Metals Corporation - Forest City, MO	Vulcan Lead Products - Milwaukee, WI
Lead Batteries - Vehicle/Marine						X	X	X	X		X			X	X		X	X	X	X	X			X	X	X	X	X	X	X	X	X	X	
Lead Batteries - Power Supply						X	X	X	X		X			X	X		X	X	X	X	X			X	X	X	X				X	X	X	
Lead Batteries - Consumer Appliances						X	X		X		X			X	X					X	X			X	X	X	X				X	X	X	
Lead Batteries - Industrial/Governmental						X	X		X		X			X	X					X	X			X	X	X	X							
Sulfuric Acid from Batteries								X				X								X				X	X									
Lead Baghouse Filtrate									X	X	X	X		X	X		X			X				X	X	X	X				X	X	X	X
Cable Sheathing		X			X	X	X	X	X	X	X		X	X	X	X	X	X	X	X	X	X		X	X	X	X				X	X	X	
Lead Paint Removal Debris			X						X			X		X	X									X	X	X	X					X	X	
Solder Drosses & Scrap	X		X		X	X	X	X	X			X	X	X	X	X	X		X	X	X		X	X	X	X	X				X	X	X	
Counter Weights	X	X			X	X	X	X	X	X	X		X	X	X	X	X	X	X	X	X	X		X	X	X	X				X	X	X	X
Plating or Chemical Solution/Sludges			X						X			X											X	X		X	X				X			
Scrap Ballast	X	X		X	X	X	X	X	X	X	X		X	X	X	X	X	X	X	X	X	X		X	X	X	X				X	X	X	X
Firing Range Soils		X	X						X			X		X	X					X				X	X	X	X							
Babbitt/"Type" Metals	X				X	X	X	X	X		X		X	X	X	X	X		X	X	X	X	X	X	X	X	X				X	X	X	X
Lead Pipe/Fittings/Coils	X	X			X	X	X	X	X	X	X			X	X	X	X	X	X	X	X	X		X	X	X	X				X	X	X	X
Superfund-type Soils			X					X	X			X												X		X	X							
Lead Solder/Paste					X	X	X	X	X	X		X	X	X	X	X	X		X	X	X		X	X	X	X	X				X	X	X	X
Lead Crystal/Glass Cullet			X					X	X	X	X	X		X	X		X							X	X	X	X					X	X	
Wheel Weights	X				X	X	X	X	X	X	X		X	X	X		X	X	X	X	X	X		X	X	X	X	X	X	X	X	X	X	X
Printed Circuit Boards									X							X					X		X	X		X	X							
Lead Sheeting/Flashing/Came	X	X		X	X	X	X	X	X	X	X		X	X	X	X	X	X	X	X	X	X		X	X	X	X				X	X	X	X
Lead Slag			X					X	X			X		X	X									X	X	X	X				X	X	X	
Radiation Shielding Materials	X	X		X	X	X	X		X	X			X	X	X	X	X		X	X		X		X	X	X	X				X	X	X	X
Lead-bearing Ceramic Scrap			X					X	X	X		X							X		X			X	X	X	X					X	X	
Lead Anodes	X	X		X	X	X	X	X	X	X			X	X	X	X		X	X	X	X		X	X	X	X	X	X	X	X	X	X	X	X
Casting Scrap	X				X	X	X	X	X		X		X	X	X	X	X		X	X	X	X	X	X	X	X	X				X	X	X	X
Lead Pigments			X						X			X		X	X									X	X	X	X					X	X	
Spent Lead Ammunition from Firing Ranges						X	X	X	X	X	X	X		X	X					X	X			X	X							X	X	
Lead-bearing Superfund Cleanup Materials/Wastes								X	X			X		X	X									X	X							X	X	

IV. HEALTH EFFECTS OF LEAD AND LEAD COMPOUNDS

A. Recent Research on Health Effects

Since the first edition of the Lead Regulation Handbook, researchers at medical centers, universities and government organizations in many nations of the world have published numerous articles on the health effects of lead and lead compounds and related subjects.

The articles on health effects added extensive information about health effects already known. The articles also provided information about possible health effects which previously received limited attention such as cancer, periodontal bone loss, obesity, asthma and schizophrenia, and attention deficit hyperactive disorder.

The articles on related subjects included subjects such as the effectiveness of abatement methods and educational programs, hazards created by demolition of buildings, and hazards created by removal methods such as dry abrasive sand blasting.

Some of the articles make difficult reading because they presume knowledge of chemistry, analytical methods and equipment, toxicology, medicine, and even dentistry. Some articles summarize the methods and results of large complex studies conducted over many years. Some articles discuss disagreements which some researchers have with methods and conclusions of other researchers.

A striking feature of the recent articles is the extent of international participation by researchers in nations such as Australia, Brazil, Denmark, Greenland, Israel, Korea, Mexico, Nigeria, Russia, Sweden, Taiwan and the United Kingdom.

Later in this Chapter, readers will find discussions of many of the leading articles together with citations to scientific journals where they can read the articles in full.

B. Historical Observations

Acute adverse effects of excessive exposure to lead have been known for centuries. Over 2,000 years ago, Hippocrates and Nicander, physicians in Greece, described the effects of severe exposures as did Pliny during the era of the Roman Empire. Some authors have written interesting papers about the possible effects of lead exposure on ancient civilizations such as Peru and Rome.[1] One article linked saturnine gout, wine consumption during the Roman

---

[1] "Lead and Silver in the Ancient Aegean," Scientific American, Gale and Stos-Gale, 244(6), pp. 176-192 (June 1981); "Contaminated and Natural Lead in Environments of Man," Arch. Environmental Health, Patterson, Vol. II,

Empire and reports of the poor health and faults of several Roman Emperors.[2] The limitations of such efforts to reconstruct ancient history as a basis for modern decisions was described in a 1994 article.[3]

Severe effects of lead exposure were described in 1700 by Bernadino Ramazzini based on observations of pottery workers and in 1786 by Benjamin Franklin who had worked as a printer.[4]

The extensive use of lead in pewter, medicines and other food containers and in the storage and transportation of water resulted in high lead exposure from the middle ages until late in the last century. Bone and hair studies show levels over 100 micrograms per gram for about 1000 until 1850. After 1850, they sharply declined to about 25 micrograms per gram as lead in the human diet was reduced by reductions in the use of pewter, lead-containing medicines and water piping. Interestingly, this decline occurred during a period when industrial use of lead greatly increased.

C. Development of Health Research and Care Industries; Health Studies

In recent decades, there have been major advances in industry, employment stability and general living conditions including health among workers and the general population. There has also been extensive development of scientific instrumentation and methods. Both industry and government have devoted major resources to health research. New drugs, surgical techniques and other corrective methods have received the most attention. However, major advances have also been made in preventive methods, beginning with

---

pp. 344-360 (1965); "Lead Poisoning and the Fall of Rome," Journal of Occupational Medicine, Gilfillan, Vol. 7, pp. 53-60 (1965)

[2] "Saturnine Gout Among Roman Aristocrats - Did Lead Poisoning Contribute to the Fall of the Empire," New England Journal of Medicine, Nriagu, Vol. 308, No. 11, pp. 660-663 (1983). Readers who enjoyed "I, Claudius," as skillfully portrayed by the actor Derek Jacoby, may be interested in the health problems of Roman nobility described by Dr. Nriagu.

[3] "The Posthumous Papers of Leaded Gasoline," 21st Century Science & Technology, Jaworowsky, Vol. 7, No. 1, Spring 1994.

[4] Letter from Benjamin Franklin to B. Vaughan, July 31, 1986.

identification of health problems and their causes and following with development of methods to prevent or control them.

Major advances have also been made in the development of synthetic materials designed to have useful characteristics without some or all of the adverse characteristics of the materials they replace. Thus, synthetic materials have reduced the need to rely on lead and other useful natural materials with adverse side-effects.

Health effects of lead have been studied with increasing frequency by health research organizations supported by funding from governmental agencies, universities and foundations. The initial focus was on workplaces where exposures tended to be greater than the general population and preventive measures could be implemented. During recent decades, the studies have been expanded to larger populations (including whole industries, localities, regions and nations) by using epidemiological and other methods. As discussed later in more detail, epidemiological studies produce results which are useful, but the conclusions that can be based on individual studies are uncertain because each study is limited by the information available, the methods used, confounding factors, and the need to make assumptions and interpretations.

During recent decades, the growth of the relatively new field of neuropsychology has played a major role in providing associative information about cognitive and behavioral effects of lead in very young children, at subsymptomatic levels, although causative mechanisms remain difficult to identify by traditional medical methods.

D. Lead Exposure Sources

Occupational exposures to lead are primarily attributable to airborne dust and fumes which may be inhaled by workers in the absence of fully effective engineering controls and personal protective equipment. However, workers can also ingest lead dust and fumes as a side-effect of inhalation and through hand-to-mouth actions such as eating or smoking without washing thoroughly.

The general population is exposed to lead from several sources. Air emissions are an important source, especially for persons residing near facilities where lead is manufactured, processed or used. However, EPA's statistics show major reductions of both total emissions and concentrations in recent decades. Dust is another important source and results from the historical use of leaded gasoline which deposited vapors on the soil and also from deteriorating lead-based paint in old urban areas. Very young children are specially susceptible to lead in household dust and soil due to hand to mouth activity. Food and food containers have been greatly reduced as lead sources in the United States and other developed nations. However, in some areas of the world, food continues to be a major lead source because of plant uptake of lead in soil and continued use of containers with lead ceramics and solders. Although there is very little lead in the rivers, lakes and groundwater from which public water supplies are obtained, drinking water continues to be a lead source where old lead piping and solder continues to be used in transportation and storage of water.

Exposure to lead has been declining rapidly. For example, in 1940, Kehoe et al estimated that young adult males ingested 200 ug/day of lead from food. The U.S. Food and Drug Administration (FDA) estimated their intake at 90 ug/day for 1974 and at 10 ug/day for 1988. Blood lead levels have also declined greatly. In its Draft Toxicological Profile for Lead dated September 2005, the Agency for Toxic Substances and Disease Registry (ATSDR) said that the CDC's NHANES results indicated that from 1976 to 1991 the mean blood lead levels of the U.S. population aged 1 to 74 years dropped 78% from 12.8 to 2.8 ug/dL. For the period 1999-2002, CDC showed a further decline to 1.6 ug/dL.

E. Lead Absorption and Excretion by the Human Body

The human body inhales and ingests lead. It also exhales and excretes lead, but absorbs some of the lead before doing so. Absorption varies with the form of the lead or lead compound, the particle, size, and dietary factors. Eating regular meals with a good mineral and fiber content reduces the absorption of lead. Significant use of alcohol, especially any use which causes kidney damage, increases absorption.

Inhalation is by far the largest intake method, about 90% for the overall population. Ingestion of food and drinking water accounts for about 10%, but can be higher for children. Large particles of lead are not easily inhaled and some may be captured in mucous and swallowed. Very small particles can be inhaled and exhaled. However, some lead is inhaled and is retained in the lungs. This lead passes into the bloodstream.

Lead compounds, such as lead carbonate, are absorbed more readily than lead itself. Dermal absorption of lead is very low, but human skin may absorb organic lead compounds.

Almost all lead absorbed in blood is linked to red blood cells. Only about 3% is linked to plasma. The lead in blood travels throughout the body and primarily into bone, but also at lower levels into tissues. When exposure ceases, the migrations into blood and tissues reverse fairly rapidly unless tissues have been seriously injured by high exposure. Migrations into bone also reverse, although more slowly. Thus, blood lead levels decline when exposure ceases. For example, occasional cases when a worker's blood lead level does not decline when removed from a known exposure source may be solved by identifying a less obvious source such as sporting or other personal activities involving lead exposure or migration of accumulated lead from bone into blood.

The human body excretes lead quite readily. The major excretion routes are the gastrointestinal system and the kidneys. Bile, hair, sweat and nails are minor. Excretion increases greatly as blood lead levels rise and a greater share is excreted through the kidneys. The ATSDR has stated that a majority of the lead taken into the body of an adult will leave within a few weeks and eventually 99% will depart. However, young children may retain over 30%. Under continued exposure, lead will accumulate in body tissues, especially bone.

F. Health Effects

Several metals (including iron, calcium and zinc) are found naturally in the human body and perform useful functions. Lead is not naturally contained, and does not perform useful functions, in the human body.

1. Acute Effects and Symptoms

Historically, the symptoms of acute lead poisoning were colic, wrist drop and gum lead lines. Gout was also associated with lead poisoning, although it also had other causes. These historical symptoms are rarely now seen in nations such as the United States.

Other symptoms that may appear in adults at blood lead levels of 80 to 100 ug/dL, but lower for some susceptible persons, including headaches, stomach aches, fatigue, muscular weakness, bowel irregularity, sleeplessness, hypertension and behavioral problems. These symptoms also result from many other causes, making them of limited use in medical diagnosis. Thus, testing for blood lead level, ZPP, CPP and ALA (defined later) is important to diagnosis. Experienced medical experts caution general practitioners not to base a diagnosis solely on possible exposure, nonspecific symptoms and testing that is inadequate.

The same symptoms may occur in children at lower levels. Differences of opinion among health professionals as to effects on children and the levels at which they occur are described later in this Chapter.

In 2005, medical and dental university researchers in Brazil published an article discussing the advantages and limitations of the use of blood lead concentrations as the primary biomarker to measure exposure to lead. The article compares the advantages and limitations of other biomarkers, i.e., plasma/serum lead, saliva lead, hair lead, urinary and fecal lead, nail lead, bone lead, and tooth lead. "A Critical Review of Biomarkers Used for Monitoring Human Exposure to Lead: Advantages, Limitations and Future Needs," Barbosa et al., Environmental Health Perspectives, Vol. 113, No. 12, p. 1669, Dec. 2005. See also "The Epidemiology of Lead Toxicity in Adults: Measuring Dose and Consideration of Other Methodological Issues," Hu et al., Environmental Health Perspectives, Vol. 115, No. 3, p. 455, March 2007.

2. Hematological Effects

An important harmful effect of lead is to inhibit cellular production of heme which carries oxygen in red blood cells. Heme normally contains iron. Lead may substitute itself for the iron. The lead inhibition results in increased production of zinc protoporphyrin (ZPP), coproporphyrins (CPP) and aminolevulinic acid (ALA). Effects of lead exposure can be measured by testing for ZPP, CPP and ALA. ZPP, which is the same as free erythrocyte protoporphyrin (FEP), resides in cells for their lifetime which is about 120 days. CPP and ALA increase and decline more rapidly than ZPP when acute exposures occur and cease. Performance of all three tests can be helpful in diagnosis.

Measures that are more sensitive at lower levels are ALA (ALAD) dehydratase to identify enzyme inhibition in blood and pyrimidine 5' nucleotidase enzyme activity (Py-5'-N).

Increase of blood lead level and inhibition of heme synthesis, as evidenced by production of ZPP, CPP and ALA, are early measurable effects that may result from lead exposure. They occur at lower levels than anemia and neurological, renal and reproductive effects which are more serious. The "action level" set by the Occupational Safety and Health Administration (OSHA) at 30 ug/100 gm is below the levels at which significant reversible effects occur such as anemia, brain and nerve damage, kidney damage and adverse pre-natal and post-natal effects in otherwise healthy adults. While studies have produced different data interpretations, these effects appear to commence for adults at levels ranging from 70 ug/100 gm to 100 ug/100 gm.

In 2004, university medical researchers in Korea published an article on a cross-sectional study of Korean 1,219 male battery and lead production workers to determine the effect of ALAD isozymes on blood lead. The article said that lead inhibits three enzymes—ALAD, coporphyrinogen oxidase, and ferrochelatase—in the heme synthesis pathway, with the effects on ALAD being the most profound. The article said the study confirmed the protective effects of ALAD 1-2 and 2-2 polymorphisms against blood lead on hematological pathways.[5]

3. Cardiovascular Disease

Blood lead levels have long been associated with increased risk of cardiovascular disease. In 2006, researchers at universities and medical centers in the United States published an article under the initial name Andy Menke, describing a study using data from NHANES III to measure the association between deaths from cardiovascular disease and blood lead levels including levels below 10 ug/dL. The researchers found an association at levels as low as 2 ug/dL, although risk of death from strokes and other causes was higher.[6]

4. Hypertension (High Blood Pressure)

Lead exposure has long been linked to hypertension in laboratory animals and adult workers in industries which produce or use lead or lead compounds. However, a recent

---

[5] "The Protective Effect of δ-Aminolevulinic Acid Dehydratase 1-2 and 2-2 Isozymes against Blood Lead with Higher Hemotological Parometers," Kim et al., Environmental Health Perspectives, Vol. 112, No. 5, April 2004.

[6] "Blood Lead Below 0.48 umol/L (10 ug/dL) and Mortality Among U.S. Adults," Menke et al., Circulation, Vol. 114, Issue 13, p. 1388, Sept. 2006. "Lead Exposure and Cardiovascular Disease--A Systematic Review," Navas—Acien et al., Environmental Health Perspectives, Vol. 115, No. 3, p. 472, March 2007.

five-year study of 780 lead-exposed children surprisingly reported no significant indication that lead raises blood pressure in children.[7]

5. Neurological Effects

Before and early in the last century, prolonged excessive exposures to lead resulted in brain disease and muscular deterioration among workers.[8] Such exposures have been a rarity in recent decades in places where modern safety methods are used. However, numerous epidemiological studies and surveys during the last two decades have raised concern about neurological effects of low-level lead exposure on very young children and recently on somewhat older children. These studies and surveys are discussed later in this Chapter.

6. Renal Effects

Since the kidneys are the organs involved in the excretion of most of the lead ingested into the human body, they can be affected by continued excessive lead exposure. Renal disfunction is usually reversible unless tissue is seriously damaged. However, prolonged excessive exposure can result in irreversible damage.

In 2003, researchers at universities and medical centers in the United States and Korea published an article describing the results of a cross-sectional study of participants in the U.S. Department of Veterans Affairs Normative Aging Study in which ALAD polymorphism status was available for 709 subjects.[9] The study investigated whether a known ALAD exon 4 polymorphism has a modifying effect on the association of blood or bone lead level with uricemia and renal function among middle-aged and elderly men. The study produced a variety of results, some of which were significant and others of which were not. The researchers

---

[7] "The Effect of Chelation on Blood Pressure in Lead-Exposed Children: A Randomized Study," Chen et al., Environmental Health Perspectives, Vol. 114, No. 4, April 2006. See also the review by Cynthia Washam on page A241 of the same issue.

[8] There is an interesting article on a recent study of 74 employees of a primary lead smelter at New Brunswick, Canada. "Effect of Lead Exposure and Ergonomic Stressors on Peripheral Nerve Function," Blecker et al., Environmental Health Perspectives, Vol. 113, No. 12, Dec. 2005. See also "Cumulative Lead Dose and Cognitive Function in Adults," Shih et al., Environmental Health Perspectives, Vol. 115, No. 3, March 2007.

[9] "The δ-Aminolevulinic Acid Dehydratase (ALAD) Polymorphism May Modify the Relationship of Low-Level Lead Exposure to Uricemia and Renal Function: The Normative Aging Study," Wu et al., Environmental Health Perspectives, Vol. 111, No. 3 p. 335, March 2003.

concluded that their findings suggested that ALAD genotype may modify the effect of lead on renal excretion of uric acid as well as overall renal function among middle-aged and elderly men who had community (nonoccupational) exposures to lead.

In 2003, researchers at universities and medical centers in the United States and Korea published an article describing the results of a study of 798 present and former lead workers in Korea to determine possible associations of renal function with polymorphisms in ALAD, vitamin D receptor (VDR) and endothelial nitric oxide synthase (e NOS) genes.[10] They found, among other things, that high lead dose was associated with higher creatinine clearances in participants with the ALAD 1-2 genotype.

In 2004, researchers from universities and medical centers in the United States published an article describing the results of a prospective study of changes in renal function during six years of followup in relation to baseline lead levels, diabetes, and hypertension among 448 middle-age and elderly men who were not occupationally exposed to lead and who were a subsample of the Normative Aging Study.[11] Again, the study produced a variety of results, some statistically significant and others not. The researchers concluded, among other things, that longitudinal decline of renal function among middle-age and elderly individuals appears to depend on both long-term lead stores and circulating lead. The effect is most pronounced among diabetics and hypertensives who likely represent particularly susceptible groups.

In 2005, researchers with universities and medical centers in the United States and South Korea published an article describing the results of their analysis of data from 803 current and former lead workers to determine whether lead biomarkers were associated with uric acid and whether previously reported associations between lead dose and renal outcomes were altered after adjustment for uric acid.[12] The study produced a variety of significant and nonsignificant

[10] "Associations of Renal Function with Polymorphisms in the δ-Aminolevulinic Acid Dehydratase, Vitamin D Receptor, and Nitric Oxide Synthase Genes in Korean Lead Workers," Weaver et al., Environmental Perspectives, Vol. 111, No. 13, p. 1613, Act. 2003.

[11] "Lead, Diabetes, Hypertension, and Renal Function: The Normative Aging Study," Tsaih et al., Environmental Health Perspectives, Vol. 112, No. 11, p. 1178, Aug. 2004.

[12] "Associations Among Lead Dose Biomakers, Uric Acid, and Renal Function in Korean Lead Workers," Weaver et al., Environmental Health Perspectives, Vol. 113, No. 11, p. 36 (Jan. 2005).

results. The researchers said that their data suggested that older workers comprise a susceptible population for increased uric acid due to lead. They added that uric acid may be one, but not the only, mechanism for lead-related nephrotoxicity.

7. Reproductive Effects

Limited studies have indicated that male lead exposure at levels greater than 40 to 50 ug/dL may result in reduced sperm counts or effects on tests.[13]

There is concern that female lead exposure may cause spontaneous abortions or miscarriages. A relatively large epidemiological study of 639 women (304 exposed, 335 unexposed) did not find an association between lead exposure and spontaneous abortion.[14]

It has been known since the middle of the nineteenth century that lead in a mother's blood can be transmitted through her placenta to a developing fetus.[15] Thus, there is concern that female lead exposure may have adverse health effects on developing fetuses that will survive birth.

In 2003, university and government researchers on the United States published an article describing an assessment using data from NHANES III to analyze the potential relation between blood level and attainment of sexual maturity by girls of ages 10 to 16 in the United States. The article reported delayed attainment of menarche and pubic hair, but not breast development.[16]

---

[13] "Reproductive Ability of Workmen Occupationally Exposed to Lead," Arch, Environmental Health, Lancrajan et al., 30:396-401 (1975). "Effects of Occupational Exposure to Lead on Sperm and Semen," Wild et al., Plenum Press, 270-300 (1983). See also "Long-Term, Low-Dose Lead Exposure Alters the Sonadotropin-Releasing Hormone System in the Male Rat," Sokol et al., Environmental Health Perspectives, Vol. 110, No. 9, p. 871, Sept. 2002.

[14] "Past Pregnancy Outcomes Among Women Living in the Vicinity of a Lead Smelter in Kosovo, Yugoslavia," Amer. Journal of Public Health, Murphy et al., Jan. 1990.

[15] "Study of the Effect of Slow Lead Intoxication on the Product of Conception," Paul, Archives of General Medicine, 15:513-533(1860). "Determinants of Elevated Blood Lead during Pregnancy in a Population Surrounding a Lead Smelter in Kosovo, Yugoslavia," Graziano et al., Environmental Health Perspectives, 1990.

[16] "Blood Lead Levels and Sexual Maturation in U.S. Girls: The Third National Health and Nutrition Examination Survey," Wu et al., Environmental Health Perspectives, Vol. 111, No. 5, p. 737, May 2003.

In 2004, researchers in Australia and the United States published an article describing a study which indicated that calcium supplementation delays increase in blood lead and reduces lead released from bone during late pregnancy and post-partum and may provide less lead exposure to a developing fetus and new born infant.[17]

Industrial programs to limit lead exposure of pregnant women workers, including a controversial policy to exclude women with childbearing capacity from working in high lead exposure jobs, were described in detail in International Union, UAW v. Johnson Controls, Inc., 886 F.2d 871 (7th Cir. 1989); reversed 499 U.S. 187 (1991). However, one epidemiological study did not find an association between cord blood lead levels at birth and later problem behaviors at age 8.[18]

8. Carcinogenicity

Studies using high doses of lead acetate and its trihydrate and of lead phosphate resulted in a conclusion by the International Agency of Research on Cancer (IARC) and the EPA that there is sufficient evidence of their carcinogenicity to experimental animals.

In its draft Public Health Statement for Lead dated September 2005, the Agency for Toxic Substances and Disease Registry (ATSDR) stated that we have no conclusive proof that lead causes cancer in humans. In 2005, the National Toxicology Program (NTP) concluded that lead and lead compounds are reasonably anticipated to be human carcinogens. The EPA has concluded that lead is a probable human carcinogen. The International Agency for Research on Cancer (IARC) has found that lead and lead compounds are probable human carcinogens.[19] ACGIH classifies lead as an animal carcinogen.

In 2002, researchers at the National Cancer Institute and the CDC published an article describing their study of the relationship of blood lead level and all cancer mortality

---

[17] "Blood Lead Changes During Pregnancy and Postpartum with Calcium Supplementation," Gulson et al., Environmental Health Perspectives, Vol. 112, No. 15, p. 1499, Nov. 2004.

[18] "Pre-and Postnatal Lead Exposure and Behavioral Problems in School-Aged Children," Environ Res., Bellinger et al. July, 1994.

[19] "IARC Carcinogen Update," Environmental Health Perspectives, Rousseau et al., Vo. 111, No. 9, Sept. 2005.

among whites in the United States using data from the National Health and Nutrition Examination Survey II (NHANES II) Mortality Study, 1992. The analyses were restricted to whites because blood lead concentrations and cancer deaths vary by race, and the number of deaths in other races in the NHANES II Mortality Study were too small to provide reliable estimates. The researchers reported that the dose response relationship using Cox proportional hazard regression models was not significant among men and women combined. Spline analyses found no dose response, and no site-specific cancer relative risks were significant. Among men, no significant dose response relationships were found for quartile or spline dose responses. For women, the results were similar except that the spline dose response results were significant at the highest levels of lead. The researchers said there was no clear biological explanation for the exception and that further replication is needed before it could be considered believable. The researchers concluded that individuals with blood levels in the range of NHANES II did not appear to have increased risk of cancer mortality.[20] An article published in 2006 by researchers who worked with NHANES III data came to a similar conclusion.[21]

9. Obesity

A study of chronic lead exposure and physical growth in children has suggested that such exposure may result in obesity.[22]

10. Asthma

A study to identify if possible increased risk of asthma could be associated with blood lead levels concluded that no significant increase was observed among either Caucasian children or African American children.[23] The African American children were at a significantly increased risk of asthma, so the authors recommended further exploration of the complex

[20] "The Association of Blood Lead Level and Cancer Mortality Among Whites in the United States," Environmental Health Perspectives, Jemal et al., Vol. 110, No. 4, April 2002.

[21] "Blood Lead Below 0.48 umol/L (10 ug/dL) and Mortality Among US. Adults," Menke et al., Circulation, Vol. 114, Issue 13, p. 1388, Sept. 2006.

[22] Environmental Health Perspectives, Kim et al., Vol. 103, No. 10, October 1995.

[23] "Blood Lead Level and Risk of Asthma," Environmental Health Perspectives, Joseph et al., Vol. 113, No. 6, p. 900, June 2005.

interrelationships between race, asthma phenotype, genetic susceptibilities, and socioenvironmental exposure, including lead.

11. Schizophrenia

In 2004, researchers at universities and a healthcare center in the United States published an article describing a possible association between prenatal exposure to lead and schizophrenia.[24] The authors discussed limitations of the study such as modest sample size, possible confounding factors, and postnatal modifying factors such as socioeconomic status. They concluded that the effects of lead may extend into later life and must be further investigated as risk factors for adult psychiatric diseases.

Later in 2004, a researcher from Johns Hopkins University sent a letter to Environmental Health Perspectives providing information supporting the possibility that prenatal lead exposure may be a risk factor for the expression of schizophrenia later in life.[25]

12. Brain Tumors

Lead is the best known of several chemicals that can inhibit ALAD. In 2005, government, university and medical center researchers in the United States published an article describing use of data from a case control study with 782 incident brain tumor cases and 799 controls that observed no increased risk for glaucoma or acoustic neuroma associated with the ALAD 2 allele. However, the findings suggested that the ALAD 2 allele may increase genetic susceptibility to mengioma. See "δ-Aminolevulinic Acid Dehydratase Polymorphism and Risk of Brain Tumors in Adults," Rajaraman et al., Environmental Health Perspectives, Vol. 113, No. 9, p. 1209, Sept. 2005.

13. Dental Problems

In 2002, dental and medical researchers in the United States published an article describing a cross-sectional analysis of data for 543 children 6-10 years old screened for enrollment in a study of potential health effects of mercury in silver fillings. The analysis

[24] "Prenatal Lead Exposure δ-Aminolevulinic Acid, and Schizophrenia," Opler et al., Environmental Health Perspectives, Vol. 112, No. 5, p.548, April 2004.

[25] "Prenatal Lead Exposure and Schizophrenia: A Plausible Neurobiologic Connection," Guilarte, Environmental Health Perspectives, Vol. 112, No. 13, p. A 724, Sept. 2004.

positively associated blood lead level with number of caries among urban children, but not with caries in a rural subgroup.[26]

In 2002, the CDC published an article discussing the relation between blood lead levels and periodontal bone loss using data from NHANES III. The researchers concluded that increased blood lead levels may be associated with advanced periodontal bone loss, particularly among people with a history of smoking.[27]

14. Attention Deficit Hyperactivity Disorder (ADHD)

In 2006, researchers with universities and medical centers in the United States published an article under the initial name Joe Braun describing a study using data from NHANES 1999-2002 to examine the association of exposures to tobacco smoke and environmental lead with attention deficit hyperactivity disorder (ADHD). The researchers reported that, using multivariable analysis, prenatal tobacco exposure and higher blood lead concentration were significantly associated with ADHD. They concluded that exposure to prenatal tobacco and environmental lead are risk factors for ADHD in U.S. children.[28]

G. Health Effects on Children

Since the publication in 1974 of an article about a pioneering study by Perino and Ernhart, numerous articles have been published describing epidemiological and other toxicological studies of neurological effects of lead exposure on children, especially effects on their intelligence and behavior. These articles led to widespread adoption of laws, regulations, policy statements and guidelines. Developments since 1990 include the following:

1. The CDC's 1991 Statement

In October 1991, the Centers for Disease Control (CDC) of the U.S. Department of Health and Human Services (DHHS) issued its Statement titled "Preventing Lead Poisoning in

[26] "Blood Lead Level and Dental Caries in School Age Children," Gemmel et al., Environmental Health Perspectives, Vol. 110, No. 10, p. A625, Oct. 2002.

[27] "The Relationship between Blood Lead Levels and Periodontal Bone Loss in the United States, 1988-1994," Dye et al., Environmental Health Perspectives, Vol. 110, No. 10, p. 997, Oct. 2002.

[28] "Exposures to Environmental Toxicants and Attention Deficit. Hyperactivity Disorder in US. Children," Braun, Environmental Health Perspectives, Vol. 114, No. 10, p. 1289, Sept. 2006.

Young Children" reducing its threshold action level for blood level content in children of age less than 6 from 25 ug/dL to 10 ug/dL. The CDC concluded that virtually all children are at risk of lead poisoning and recommended universal screening, except in communities where large numbers or percentages of children have been screened and found not to have lead poisoning. The CDC recommended that screening be accomplished by blood lead tests rather than EPP measurement which it described as not sensitive enough. The CDC provided the following Table 1-1 in Chapter 1 of the Statement:

Class	Blood Lead Concentration	Comment
I	<9	A child in Class I is not considered to be lead-poisoned.
IIA	10-14	Many children (or a large proportion of children) with blood lead levels in this range should trigger community-wide childhood lead poisoning prevention activities. Children in this range may need to be rescreened more frequently.
IIB	15-19	A child in Class IIB should receive nutritional and educational interventions and more frequent screening. If the blood lead level persists in this range, environmental investigation and intervention should be done.
III	20-44	A child in Class III should receive environmental evaluation and remediation and a medical evaluation. Such a child may need pharmacologic treatment of lead poisoning.
IV	45-69	A child in Class IV will need both medical and environmental interventions, including chelation therapy.
V	>70	A child with Class V lead poisoning is a medical emergency. Medical and environmental management must begin immediately.

Some of the members and consultants of the CDC's advisory committee listed in the 1991 Statement participated in well-known studies and surveys of the effects of lead exposure on children, which led to the CDC's conclusions.

In its website publication titled "Lead: Questions and Answers" shown on December 6, 2006, the CDC gave reasons not to lower the 10 ug/dL "levels of concern" at this time. In brief, they were lack of effective clinical interventions to further lower levels for children, inaccuracy inherent in laboratory testing, and lack of evidence of a threshold below which adverse effects are not experienced.

2. Questioning the CDC's Methods and Conclusions

In 1993, Craig Boreiko, Manager, Environmental Health, of International Lead Zinc Research Organization, Research Triangle Park, and Claire Ernhart, Professor, Department of Psychiatry and Reproductive Biology, Metro-Health Medical Center, Cleveland, questioned the CDC's methods, conclusions and recommendations. Their discussion outline for the Lead Tech '93 Conference said that the CDC had relied on research which lacked sufficient controls of confounding factors and had ignored studies and portions of studies which did not support its position. They pointed out that the CDC's conclusion of adverse effects on young children at 10 ug/dL resulted in a tenfold increase in the number of affected children, resulting in an epidemic by edict, although lead levels were actually decreasing. They questioned the effectiveness of chelation and the need to frighten parents and use medical time for tests that are seldom of any consequence.

3. The Pirkle Report (NHANES III)

In 1994, the Pirkle report from the Third National Health and Nutrition Examination Survey (NHANES III) conducted by the CDC confirmed that blood lead levels of the U.S. population have been declining substantially. Comparing the NHANES III results for 1988 to 1991 to the NHANES II results for 1976 to 1980, the mean blood lead levels for persons aged 1 to 74 years dropped 78% (12.8 to 2.8 ug/dL). Mean blood lead levels of children aged 1 to 5 years declined 77% (13.7 to 3.2 ug/dL) for non-Hispanic white children and 72% (20.2 to 5.6 ug/dL) for non-Hispanic black children. The prevalence of blood lead levels of 10 ug/dL or greater declined from 85.0% to 5.5% for non-Hispanic white children and from 97.7% to 20.6% for non-Hispanic black children. Similar declines were found in other population subgroups including results from Mexican-Americans based on comparison of NHANES III results to the Hispanic Health and Nutrition Survey (NHANES) for 1982-1984. The report said that the data indicated major progress in reducing lead exposure, but that sociodemographic factors continue to be associated with higher blood lead levels including younger age, male sex, non-Hispanic black race/ethnicity and low income level. "The Decline in Blood Lead Levels in the United States," Pirkle et al., Journal of the American Medial Association, Vol. 272, No. 4 (1994).

4. The Brody Report; CDC's Comparison of NHANES II and III

In 1994, the Brody report further analyzed the sociodemographic implications of Phase I (1988 to 1991) of NHANES III. The report concluded that the low overall blood lead levels demonstrate a major public health success. However, the report also concluded that exposure to lead at levels that may adversely affect the health of children remains a problem especially for those who are minority, urban, and from low-income families.

In a booklet dated November 1997 and titled "Screening Young Children for Lead Poisoning," the CDC reported that the geometric mean blood lead levels of children ages 1-5 in the United States declined from 15.0 for NHANES II, 1976-80 to 3.6 for NHANES III, phase one (1988-91) and 2.7 for NHANES III, phase 2 (1991-94). However, CDC estimated that 890,000 U.S. children had blood lead levels equal to or greater than 10 ug/dL.

ASTDR's Draft Toxicological Profile for Lead dated September 2005 said that the geometric mean blood lead level of children of ages 1-5 years declined to 1.9 ug/dL based on sampling data for 1999-2002.

5. The Pocock Report

In November 1994, the Pocock report described its systematic review of 26 epidemiological studies since 1979 to quantify the magnitude of the relation between full scale IQ in children aged 5 or more and their body burden of lead. The studies reviewed consisted of 5 prospective studies of birth cohorts (1,100 children), 14 cross sectional studies of blood lead (3,499 children), and 7 cross sectional studies of tooth lead (2,095 children). The report stated that synthesis of this evidence, including meta-analysis, indicated that a typical doubling of body lead burden is associated with a mean deficit in full scale IQ of around 1-2 IQ points. "A Systematic Review of the Epidemiological Evidence," Pocock et al., British Medical Journal, 309 (6963), pp. 1189-1197 (Nov. 1994). After extensive discussion, the report reached the following conclusions:

> "The public health implications of low level lead exposure in children continue to provoke widespread concern in many countries. Our systematic review of the overall evidence shows a small but potentially important deficit in full scale IQ among children with raised body lead burden. However, the inherent limitations of observational epidemiology in pinpointing the reasons for this association mean that uncertainty remains as to the real impact that lead makes on children's neuropsychological development. In the face of this doubt, the priority that should be devoted to detection and intervention on children with moderately increased blood lead, compared with other social influences on childhood development, is open to debate."

6. The Tong Articles on the Port Pirie Cohort Study

Australian and English researchers with the Port Pirie Cohort Study reported in a 1996 article that exposure to environmental lead in the first seven years of life is associated with cognitive deficits that seem to persist into later childhood.[29] They reported in a 1998 article that cognitive deficits associated with exposure to environmental lead in early childhood appear to be only partially reversed by a subsequent decline in blood level.[30]

7. The Canfield Article

In 2003, researchers from several universities and medical centers funded and assisted by the U.S. National Institute of Environmental Health Sciences (NIEHS), published in an article under the lead name Canfield, describing a cohort study of the relation between blood lead concentrations in 172 children at 6, 12, 18, 24, 30, 36, 48 and 60 months of age and their IQs at 3 and 5 years. The researchers reported that blood lead concentration was inversely and significantly associated with IQ. Of greatest interest, the change in IQ associated with a given change in lead concentration was greater for a subsample of 101 children whose maximal lead concentrations remained below 10 ug/dL. When estimated using a nonlinear model with the sample, IQ declined by 7.4 points as lifetime average blood lead concentrations increased from 1

[29] "Lifetime Exposure to Environmental Lead and Children's Intelligence at 11-13 Years: The Port Pirie Cohort Study," Tong, British Medical Journal, Vol. 312, p. 1569, June 22, 1996.

[30] "Declining Blood Levels and Changes in Cognitive Function During Childhood," Tong, Journal of the American Medical Association, Vol. 280, No. 22, 1998.

to 10 ug/dL. The authors of the Canfield study concluded that declines in IQ are greater at concentrations below 10 ug/dL than at higher concentrations. Thus, they concluded that more U.S. children than previously estimated may be adversely affected by environmental lead.[31]

8. The Koller Article

In 2004, researchers from the University of Leicester and the University of Birmingham, United Kingdom published an article under the lead name of Karin Koller reviewing the Canfield study and other recent developments on low-level lead exposure and intellectual impairment in children.[32] The article, which is quite detailed, asked several questions about the methods used in the Canfield study. The article said that IQ tests are blunt measures of neurologic status, and blood lead is at best only a crude index of lead-induced neurotoxicity. The last paragraph reads as follows:

> "It is clear that blood lead levels have fallen significantly over the last 40 years. During the 1970s, childhood blood lead concentrations of 40 ug/dL were not unusual. The available evidence suggests that mean blood lead levels are now in the range 2-4 ug/dL in the United States and much of Europe. Despite this reduction in lead exposure, it could be argued that current baseline blood lead levels continue to constitute a global public health risk, as preindustrial humans are estimated to have had 100- to 1,000-fold lower blood lead levels than the population of today (Owen and Flegal 1998). With the recent evidence demonstrating an inverse association between blood lead levels and cognitive function in children exposed to low levels of lead, there is no safety margin at existing exposures. Clearly, efforts must continue to minimize childhood exposure. However we would urge that these efforts be seen in perspective. The magnitude of the lead-IQ dose-response relationship is small on a population basis and should be set against the far greater combined effect of SES status and quality of the caregiving environment. It has been argued that, instead of "chasing after an ever-receding lead threshold," attention and funds should be focused on "the more complex

---

[31] "Intellectual Impairment in Children with Blood Lead Concentrations Below 10 ug per deciliter," Canfield et al., The New England Journal of Medicine, Vol. 348, No. 16, April 17, 2003.

[32] "Recent Developments in Low Level Lead Exposure and Intellectual Impairment in Children," Koller et al., Environmental Health Perspectives, Vol 112, No. 9, p. 987, June 2004.

> social ills that are associated with continued lead exposure in a small segment of the population" (Gee and McKay 2002). Current lead exposure accounts for a very small amount of variance in cognitive ability (1-4%), whereas covariates such as social and parenting factors account for 40% or more (Weiss 2000)."

In a letter published in June 2005 in Environmental Health Perspectives, Richard Canfield and three other authors of the Canfield study commended the authors of the Koller review and responded to several of their questions. In a letter published in the same issue, Karin Koller and the other authors of the Koller review expressed gratitude for the responses relating to confounding factors and use of the Sanford-Binet test to measure IQ.[33]

9. The Chen Article

In 2005, researchers at NIEHS and universities and schools of medicine in the United States described in an article under the initial name Chen the results of a clinical trial of 780 children of 12-33 months of age with blood lead concentrations of 20-44 ug/dL at about 2 years of age in four cities in the United States. Lead levels and IQs were later tested at ages 5 and 7. The researchers reported stronger associations between IQ levels, at ages 5 and 7, and concurrent blood lead levels, at those ages, than with blood lead levels at age 2.[34] They said the analysis supports the idea that lead exposure continues to be toxic to children as they reach school age and does not support the interpretation that all the damage is done by the time a child is 2 or 3 years of age.

In the same issue of Environmental Health Perspectives, a reviewer summarized the Chen article, specially mentioning that the association of concurrent blood lead level with IQ was stronger as the children grew older, although blood lead concentrations dropped as the children aged. The reviewer pointed out that the researchers had no data reflecting how much caregivers interacted with and stimulated the children and that most of the children were from

---

[33] Letters to the Editor, Environmental Health Perspectives, Vol. 113, No. 11, p. A16, Jan. 2005.

[34] "IQ and Blood Lead from 2 to 7 Years of Age: Are the Effects in Older Children the Residual of High Blood Lead Concentrations in 2-Year-Olds?," Chen et al., Environmental Health Perspectives, Vol. 113, No. 5, May 2005.

minority racial, public assistance, single-parent households. The reviewer commented that the conclusions of the study emphasize the importance of testing for and reducing lead contamination in the environment of children much older than typically targeted.[35]

10. The Lanphear Article on an International Pooled Analysis

In 2005, researchers from medical centers and universities in the United States, Australia and Mexico published an article, under the initial name Bruce Lanphear, describing the results of an examination of data from 1,333 children who participated in seven international population-based longitudinal cohort studies followed from birth until 5-10 years of age. The researchers concluded that environmental lead exposure in children who have minimal blood lead levels less than 7.5 ug/dL is associated with intellectual deficits.[36]

In 2006, a letter from Claire Ernhart of Case Western Reserve University was published in Environmental Health Perspectives. Ms. Ernhart said that she contributed data from her Cleveland Study to the Lanphear pooled study, but withdrew from authorship because she could not concur with the manuscript, including the inferences drawn. She described several points on which she disagreed with the article and commented that it reached conclusions intended to support policies to further reduce the already low level of childhood lead exposure.[37] In the same issue, Bruce Lanphear and four other authors of the Lanphear article answered Ernhart's letter. They said that their studies provide sufficient evidence that childhood lead exposure should be reduced even more by banning all nonessential uses of lead and further reducing the allowable levels of lead in air emissions, house dust, soil, water, and consumer products.[38]

---

35 "Challenging Assumptions about Lead and IQ," Weinhold, Environmental Health Perspectives, Vol. 113, No. 5, May 2005.

36 "Low-Level Environmental Lead Exposure and Children's Intellectual Function: An International Pooled Analysis," Lanphear et al., Environmental Health Perspectives, Vol. 113, No. 7 July 2005.

37 "Effects of Lead on IQ in Children," Ernhart, Environmental Health Perspectives, Vol. 114, No. 2, Feb. 2006.

38 "Lead and IQ in Children: Lanphear et al., Respond," Lanphear et al., Environmental Health Perspectives, Vol. 114, No. 2, Feb. 2006.

11. The Schnaas Article (A Mexico City Cohort Study)

In 2006, researchers with government health organizations and universities in Mexico published an article describing a cohort study of 175 children in Mexico City to analyze the lead effect of the cohort from pregnancy through 10 years of age on child IQ from 6 to 10 years.[39] They concluded that the third trimester of pregnancy may constitute a critical period for subsequent intellectual child development during which lead exposure can produce lasting and possible permanent effects. They added that the data suggest there is no threshold for the adverse consequences of lead. On the contrary, the largest IQ changes in the study sample were observed within the first few micrograms per deciliter of blood lead.

12. American Academy of Pediatrics Policy Statement

In 2005, the Committee on Environmental Health of the American Academy of Pediatrics published a Policy Statement titled "Lead Exposure in Children: Prevention, Detection, and Management."[40] The Abstract reads as follows:

> Fatal lead encephalopathy has disappeared and blood lead concentrations have decreased in US children, but approximately 25% still live in housing with deteriorated lead-based paint and are at risk of lead exposure with resulting cognitive impairment and other sequelae. Evidence continues to accrue that commonly encountered blood lead concentrations, even those less than 10 ug/dL, may impair cognition, and there is no threshold yet identified for this effect. Most US children are at sufficient risk that they should have their blood lead concentration measured at least once. There is now evidence-based guidance available for managing children with increased lead exposure. Housing stabilization and repair can interrupt exposure in most cases. The focus in childhood lead-poisoning policy, however, should shift from case identification and management to primary prevention, with a goal of safe housing for all children.

---

[39] "Reduced Intellectual Development in Children with Prenatal Lead Exposure," Schnaas et al., Environmental Health Perspectives, Vol. 114, No. 5, p. 791, May 2006.

[40] "Lead Exposure in Children: Prevention, Detection and Management," American Academy of Pediatrics, PEDIATRICS, Vol. 116, No. 4, Oct. 2005.

A more extensive list of articles and other publications is contained in Appendix A to this Chapter. Interested readers will also find an extensive list of studies in the references at the end of the Koller article discussed earlier.

H. Epidemiological Studies and Confounding Factors

During recent decades, numerous epidemiological and other toxicological studies, health surveys and literature surveys were performed in an effort to identify and measure the health effects of lead exposure. Some focused on population or worker groups. In recent years, many have focused on very young children of age 6 or less.

The most extensive efforts in recent years have been to identify and measure neuropathological effects on very young urban children of lead exposure from sources such as paint, dust, soil and food. The methods used and conclusions reached by these studies have been controversial.

Epidemiological studies are valuable to the health community and the public because they provide useful information that might be unobtainable by other means without years of costly effort. However, epidemiological studies are difficult to perform because of the need to control, to the extent possible, for numerous confounding factors.

Confounding factors which complicate epidemiological studies of lead exposure among adults include exposure to other chemical substances, age, smoking, alcohol consumption and body mass. In worker studies, differentiation between workplace exposure and exposure during personal activities is important but difficult to accomplish because workers have no obligation to provide personal information.

Confounding factors which complicate epidemiological studies of lead exposure of young children are numerous. They include premature birth; small size at birth; problems during delivery; genetic problems; use of drugs to preserve maternal or child health; childhood diseases; family problems including neglect, cruelty, divorce, death and abandonment; linguistic problems; and all the socioeconomic problems experienced by generations of urban children.

I. Chelation

Chelation agents have been used for many years to reduce blood lead levels, usually in urgent cases where there is a risk of death or severe adverse health effects. Chelation agents are chemical compounds that bond to lead and other metals in the body, thus aiding in their excretion so as to reduce lead concentrations more rapidly than the body's normal functions would accomplish. Examples of chelating agents used for lead are edetate disodium calcium (Ca EDTA), dimercaperol, D-Penacillamine, and meso- 2,3-dimercaptosuccinic acid (succimer).

In 2001, researchers at the NIEHS and medical centers and universities in the United States published an article on the effect of chelation therapy with succimer on neuropsychological development in children exposed to lead.[41] They concluded that treatment with succimer lowered blood lead levels, but did not improve scores on tests of cognition, behavior or psychological function in children with blood lead levels below 45 ug/dL. Since succimer is as effective as any lead chelator currently available, the authors said that chelation therapy is not indicated for children with these blood lead levels.

In 2003, researchers at universities, medical centers and the NIEHS in the United States published an article describing the effect of succimer on growth of preschool children with moderate blood lead levels. They concluded that succimer does not have a beneficial effect on growth and may have an adverse effect.[42]

In 2004, researchers at universities in the United States published an article describing the results of a rodent study in which rats were exposed to lead followed by treatment with succimer.[43] They said that succimer-induced reductions in brain lead lagged behind reductions in blood lead and were generally smaller in magnitude.

In 2006, researchers at the NIEHS and medical center and university organizations published an article which described the results of a randomized clinical trial of 780 children of 12-33 months of age (mean age, 2 years) with moderately high blood lead

---

[41] "The Effect of Chelation Therapy with Succimer on Neuropsychological Development in Children Exposed to Lead," Rogan et al., New England Journal of Medicine, Vol. 344, p. 1421.

[42] "Effect of Succimer on Growth of Preschool Children with Moderate Blood Lead Levels," Peterson et al., Environmental Health Perspectives, Vol. 112, No. 2, Feb. 2004.

[43] "Reduction in Blood Lead Overestimate Reductions in Brain Lead after Repeated Succimer Regimens in a Rodent Model of Childhood Lead Exposure," Stangle et al., Environmental Health Perspectives, Vol. 112, No. 3, March 2004.

concentrations (20-44 ug/dL).[44] Clinical centers in Baltimore, Cincinnati, Newark and Philadelphia participated in the trial. The researchers said that children in a group treated with succimer had lower blood lead concentrations, but their blood pressure did not differ from those in a placebo group. They found no association between blood lead and blood pressure. They also concluded that chelation with succimer did not change blood pressure.

In 2006, the CDC published an article in the Journal of the American Medical Association describing deaths that occurred in Texas, Pennsylvania and Oregon resulting from hypocalcemia after intravenous administration of $Na._2$ EDTA, rather than Ca EDTA, as a chelation agent to reduce blood lead levels.[45] The CDC emphasized that Ca EDTA is the only agent recommended for intravenous chelation therapy for children, but said that hospital formularies usually stock multiple chelation agents. The CDC noted that deaths associated with lead poisoning are rare and that childhood deaths caused by cardiac arrest associated with chelation therapy had not been documented previously. The CDC added that, as blood lead levels in children continue to decline, fewer children require chelation therapy. However, primary care providers should consult experts in the chemotherapy of lead before using chelation drug therapy or contact state or local childhood lead poisoning programs or the CDC.

J. Studies Relating to Cleanup Needs and Methods

Researchers at HUD and NIEHS published the results of a survey between 1998 and 2000 to estimate the number of housing units in the United States with lead-based paint and lead-based paint hazards. Based on a random sample of 831 housing units between 1998 and 2000, they estimated that 38 million housing units had lead-based paint, down from 64 million in 1990, and 24 million units had significant lead-based paint hazards. The study also analyzed the prevalence of lead-based paint and hazards by factors such as household income, age of housing, and region of the country. Among other things, the study concluded that most painted surfaces,

---

[44] "The Effect of Chelation on Blood Pressure in Lead-Exposed Children: A Randomized Study," Chen et al., Environmental Health Perspectives, Vol. 114, No. 4, April 2006. However, see a later article on cognitive effects of succimer chelation by Stangle et al., Environmental Health Perspectives, Vol. 115, No. 2, p. 201, February 2007.

[45] "Deaths Associated with Hypocalcemia from Chelation Therapy – Texas, Pennsylvania and Oregon, 2003-2005," Journal of the American Medical Association, Vol. 295, No. 18, May 10, 2006.

even in older housing, do not have lead-based paint which is found on 2% to 25% of painted building components. The greatest risk is in older units with lead-based paint hazards that are or will be occupied by children under age 6 and are low income and/or undergoing renovation or maintenance that disturbs lead-based paint.[46]

In September 2001, researchers at Xavier University of Louisiana and Colorado State University published the results of a study of two old houses in New Orleans to estimate the quantities of dust containing lead and other metallic contaminants generated by power sanding and paint scraping. The study estimated that power sanding generates high volumes of metallic dusts that create severe risks, especially to children and pets. The authors recommended that all sanding be prohibited.[47]

In 2002, researchers at Howard School of Public Health, the University of Cincinnati, Johns Hopkins University, the University of Pennsylvania, Children's Hospital of Pennsylvania, NIEHS, and the University of Medicine and Dentistry of New Jersey published an article describing the results of a project to clean the homes of 765 children selected at random in order to conduct a clinical trial of the effects of oral chelation. The article reported that a single professional cleaning made substantial reductions of dust lead loadings, but did not reduce loadings of all dust samples below federal standards. The authors concluded, among things, that repeated cleaning and more aggressive lead hazard reduction strategies are required and that interim control measures are no substitute for abatement and other long-term strategies to remove sources of lead exposure.[48]

In 2002, researchers at the University of Medicine and Dentistry of New Jersey, Harvard School of Public Health, Nankai University, Taianjin, China, the University of

---

[46] "The Prevalence of Lead-Based Paint Hazards in U.S. Housing," Jacobs et al., Environmental Health Perspectives, Vol. 110, No. 10, Oct. 2002.

[47] "Multiple Metal Contamination from House Paints: Consequences of Power Sanding and Scraping in New Orleans," Mielke et al., Environmental Health Perspectives, Vol. 109, No. 9, Sept. 2001.

[48] "Assessment of Cleaning to Control Lead Dust of Children with Moderate Lead Poisoning," Ettinger et al., Environmental Health Perspectives, Vol. 110, Dec. 2002.

Minnesota School of Public Health, and HUD published an article describing the results of a randomized field study to determine whether a conventional vacuum cleaner could achieve results comparable to a high efficiency particulate air (HEPA) vacuum cleaner in cleaning lead-contaminated house dust. The researchers reported that both vacuum cleaners significantly reduced relatively high dust lead loadings from carpets. The HEPA vacuum achieved a greater reduction, but the difference was not statistically significant. The non-HEPA vacuum cleaner did not perhaps because of low precleaning dust levels. The researchers concluded that a non-HEPA vacuum cleaner can be used to clean dust in carpets if a HEPA vacuum cleaner is not available. They also encouraged frequent cleaning of carpets and upholstery because the effectiveness of one-time cleaning is limited.[49]

In 2003, researchers at Johns Hopkins University and HUD published an article on a longitudinal study to assess changes in ambient dust levels associated with demolition of three sites containing blocks of older lead-containing row houses in Baltimore, Maryland. The researchers reported high increases in lead dust fall rates and increases of the lead concentrations of lead in dust falls during demolition and during removal of debris. They recommended, among other things, improved work practices during demolition and debris removal, limitation of access to demolition sites, minimization and containment of lead dust, and advance notification to residents in the vicinity and information on steps they can take to minimize risk of lead exposure.[50]

In 2005, researchers from the University of California and the Children's Hospital of Philadelphia published an article describing the results of an isotopic estimate of bone lead contribution to blood in children, including a discussion of its implications for assessing the efficiency of lead abatement. Based on a limited number of cases, the researchers said that the

---

[49] "Comparison of Techniques to Reduce Residential Lead Dust in Carpet and Upholstery," Yiin et al., Environmental Health Perspectives, Vol. 110, No. 12, Dec. 2002.

[50] "A Study of Urban Housing Demolition as Sources of Lead in Ambient Dust," Farfel et al., Environmental Health Perspectives, Vol, 111, No. 9, July 2003.

results supported the hypothesis that release of bone lead into blood may substantially buffer the decrease in blood lead levels expected from reduction of lead intake after lead abatement.[51]

K. International Studies

In recent years, a number of articles have been published describing studies of blood lead levels in populations outside of the United States. Further, research groups at medical centers and universities outside the United States have conducted, or participated in, studies and reviews of the health effects of lead. Examples of such articles are as follows:

Australia. "Lifetime Exposure to Environmental Lead and Children's Intelligence at 11-13 Years: the Port Pirie Cohort Study," BMJ Learning.com, Vol. 312, p. 1569, 1996.

Australia. "Declining Blood Lead Levels and Changes in Cognitive Function During Childhood," Journal of the American Medical Associations, Vol. 280 p. 1915, Dec. 9, 1998.

Russia. "Lead Poisoning Among Young Children in Russia," Rubin et al., Environmental Health Perspectives, Vol. 110, No. 6 June 2002.

Sweden. "Cadmium and Lead in Blood in Relation to Low Bone Mineral Density and Low Tubular Proteinuria," Alfuen et al., Environmental Health Perspectives, Vol 110, No. 7, July 2002.

Mexico. "The Relationship Between Lead in Plasma and Whole Blood in Women," Smith et al., Environmental Health Perspectives, Vol. 110, No. 3, March 2002.

Mexico. "Relationship of Blood and Bone Lead to Menopause and Bone Mineral Density Among Middle-Age Women in Mexico City," Latorre et al., Environmental Health Perspectives, Vol. 111, No. 4, April 2003.

Sweden. "Yearly Measurements of Blood Lead in Swedish Children since 1978: An Update Focussing on the Petrol Lead Free Period 1995-2001," Stromberg et al. Occupational Environmental Medicine, Vol. 60, p. 370, 2003.

Taiwan. "Association of Renal Function with Polymorphisms in the δ-Aminolevulinic Acid Dehydratase (ALAD) Polymorphism May Modify the Relationship of Low Lead Exposure

[51] "A Noninvasive Isotopic Approach to Estimate the Bone Lead Contribution to Blood in Children: Implications for Assessing the Efficacy of Lead Abatement," Guiazda et al., Environmental Health Perspectives, Vol. 113, No. 1 p. 104, Jan. 2005.

to Uricemia and Renal Function," Wu et al., Environmental Health Perspectives, Vol. 111, No. 3, March 2003.

Korea. "Association of Renal Function with Polymorphisms in the δ-Aminolevulinic Acid Dehydratase, Vitamin D Receptor, and Nitric Oxide Synthase Genes in Korean Lead Workers," Weaver et al., Environmental Health Perspectives, Vol. 111, No. 13, Oct. 2003.

United Kingdom. "Recent Developments in Low-Level Lead Exposure and Intellectual Development in Children," Koller et al., Environmental Health Perspectives, Vol. 112, No. 9, June, 2004.

Mexico. "Blood Lead Similar Trend in a Cohort of Children in Mexico City (1987-2002), Schnaas et al., Environmental Health Perspectives, Vol. 112, No. 10, July 2004.

Greenland and Denmark. "Lead Sources in Human List in Greenland," Bjerregaard et al., Environmental Health Perspectives, Vol. 112, No. 15, Nov. 2004.

Australia. "Blood Lead Changes during Pregnancy and Postpartum with Calcium Supplementation," Gulson et al., Environmental Health Perspectives, Vol. 112, No. 15, Nov. 2004.

Korea. " Associations among Lead Dose Biomarkers, Uric Acid, and Renal Function in Korean Lead Workers," Weaver et al., Environmental Health Perspectives, Vol. 113, No. 1, Jan. 2005.

Australia and Mexico. "Low-Level Environmental Lead Exposure and Children's Intellectual Function: An International Pooled Analysis," Lanphear et al., Environmental Health Perspectives, Vol. 113, No. 7, July 2005.

Canada. "History of Inuit Community Exposure to Lead, Cadmium, and Mercury in Sewage Lake Sediments," Hermanson et al., Environmental Health Perspectives, Vol. 113, No. 16, Oct. 2005.

Nigeria. "Lead Contamination in Cocoa and Cocoa Products: Isotopic Evidence of Global Contamination," Environmental Health Perspectives, Vol. 113, No. 10, Oct. 2005.

Brazil. "A Critical Review of Biomarkers Used for Monitoring Human Exposure to Lead: Advantages, Limitations, and Future Needs," Barbosa et al., Environmental Health Perspectives, Vol. 113, No. 12, Dec. 2005.

Taiwan. "Lead Increases Lipopolysaccharide-Induced Liver Injury Through Tumor Necrosis Factor-oc Overexpression by Monocyte/Microphages," Chen et al., Environmental Health Perspectives, Vol. 114, No. 4, April 2006.

Palestinian Authority, Israel and Jordan. "Childhood Lead Exposure in the Palestinian Authority, Israel and Jordan: Results from the Middle Eastern Regional Cooperation Project, 1996-2000, Safi et al., Environmental Health Perspectives, Vol. 114, No. 6, June 2006.

Taiwan. "Lead Exposure Is Associated with Decreased Serum Paroxonase 1 (PON 1) Activity and Genotypes," Li et al., Environmental Health Perspectives, Vol. 114, No. 8, Aug. 2006.

NOTE: In March 2007, faculty members of several major medical and public health schools and institutes in the United States and Mexico published a mini-monograph in Environmental Health Perspectives, Vol. 115, No. 3, p. 451, March 2007. The mini-monograph includes five articles which review existing literature on the health effects of lead on adults. The articles, which do not themselves present new scientific data, begin with an article calling for stricter occupational health standards to protect lead-exposed adults in the United States and around the world. The associated articles discuss the biomarkers used to measure lead dosage; medical management of adult lead exposure; cardiovascular disease; and cumulative dose effects on cognitive function in adults. The introductory article recommends that OSHA modify its lead standards, which protect workers against both acute health effects of recent dose and chronic health effects of cumulative dose, i.e., (1) <20 ug/dL to protect against acute effects and (2) tibia lead levels not exceeding 15 ug lead/g bone mineral, or a cumulative blood lead level index below approximately 200-400 ug-years/dL, to protect against chronic effects. The authors explained that the cumulative blood lead level index is equivalent to an average blood lead level of 20 ug/dL for 10-20 years or of 10 ug/dL for 20-40 years.

Appendix A,
Chapter IV

EPIDEMIOLOGICAL STUDIES, HEALTH SURVEYS AND REVIEWS OF NEUROLOGICAL EFFECTS OF LOW-LEVEL LEAD EXPOSURE ON YOUNG CHILDREN

The extent to which low-level lead exposure may have effects on children up to age 6, and more recently up to early teen age, is controversial. The controversy includes scientific, political, economic and legal elements. Accordingly, the author decided to provide a list of published articles which illustrate the different points of view:

- "The relation of subclinical lead level to cognitive and sensorimotor impairment to black preschoolers," Perino and Ernhart, Journal of Learning Disorders, Vol. 7, pp. 708-712 (1974).
- "Neurolopsychological dysfunction in children with chronic low-level lead absorption," Landrigan et al., Lancet, Vol. 1, pp. 708-712 (1975)
- "Deficits in Psychological and Classroom Performance of Children with Elevated Dentine Lead Levels," Needleman, et al., New England Journal of Medicine, Vol. 300, No. 13, pp. 689-731 (1979).
- "Subclinical Levels of Lead and Developmental Deficit - a Multivariate Follow-up Reassessment," Ernhart et at., Pediatrics, Vol. 67, pp. 911-919 (1981)
- "Influence of Blood Lead on the Ability and Attainment of Children in Edinburgh," Fulton et al, Lancet, pp. 1221-1225 (May 30, 1987).
- "Port Pirie Cohort Study: Environmental Exposure to Lead and Children's Abilities at the Age of Four Years," McMichael et al., New England Journal of Medicine, Vol 319, No. 8, pp. 468-475 (1988).
- "Low Level Lead Exposure and Intelligence in the Preschool Years," Ernhart et al., The Science of the Total Environment, 71:453-459 (1988).
- "Prenatal and Preschool Age Lead Exposure: Relationship with Size," Neurotoxicology and Teratology, 13(4) 1-11 (1991).

- "Low Level Lead Exposure in the Prenatal and Early Preschool Periods: Intelligence Prior to School Entry," Ernhart et at., Journal of Clinical Epidemiology (1993).
- "Environmental Exposure to Lead and Children's Intelligence at the Age of Seven Years - The Port Pirie Cohort Study," Baghurst et al., New England Journal of Medicine, Vol. 327, pp. 1279-1284 (1992).
- "Dentine Lead and Intelligence Prior to School Entry: A Statistical Sensitivity Analysis," Greene and Ernhart, Journal of Clinical Epidemiology (1993).
- "A Critical Review of the Pre-Natal Lead Exposure in the Human - Effects on the Developing Child," Ernhart, Reproduction Toxicology, Vol. 6, pp. 21-40 (1992).
- "Low-Level Lead Exposure, Intelligence and Academic Achievement: A Long-Term Follow-up Study," Bellinger et al., Pediatrics, Vol. 90, No. 6, (1992).
- "Independent Effects of Lead Exposure and Iron Deficiency Anemia on Developmental Outcomes at age 2 Years," Wasserman et al., Journal of Pediatrics, Nov. 1992 (The Kosovo Study).
- "The Developmental Consequences of Low to Moderate Prenatal and Postnatal Lead Exposure: Intellectual Attainment in the Cincinnati Lead Study Cohort Following School Entry," Neurotoxicological Teratology, Vol. 15, pp. 37-44 (1993).
- "Pre and Post Natal Lead Exposures and Behavior Problems in School-Aged Children," Bellinger et al., Environ Res., July: 66(1), pp.12-30 (1994)
- "National Estimates of Blood Lead Levels: United States, 1976-1980 (NHANES II)," Mehaffey et al., New Environmental Journal of Medicine, Vol. 307, pp. 573-579 (1982).
- "Blood Lead Levels of 4-11 Year Old Mexican American, Puerto Rican and Cuban Children (NHANES II)," Carter-Pokras et al., Public Health Rep., Vol. 105, pp. 388-393 (1990).

- "Preventing Lead Poisoning in Young Children: A Statement by the Centers for Disease Control," Centers for Disease Control, U.S. Department of Health and Human Services, 1991.
- "Adverse Health Effects of Lead at Low Exposure Levels: Trends in the Management of Childhood Lead Poisoning," John F. Rosen, Toxicology, Elsevier (February 20, 1994).
- "Blood Leads Levels in the U.S. Population (NHANES III)" Brody et al., Journal of the American Medical Association, Vol. 272, No. 4, pp. 277-283 (1994).
- "The Decline in Blood Lead Levels in the United States, (NHANES III)" Pirkle et al., Journal of the American Medical Association, Vol. 272, No. 4 (1994).
- "Low-Level Lead Exposure and Cognitive Function in Children," Bellinger et al., Pediatric Annals, 23:11 (Nov. 1994).
- "Environmental Lead and Children's Intelligence: A Systematic Review of the Epidemiological Evidence," Pocock et al., British Medical Journal, 309 (6963), pp. 1189-1197 (Nov. 1994).
- "Lifetime Exposure to Environmental and Children's Intelligence at 11-13 Years: the Port Pirie Cohort Study," Tong et al., British Medical Journal, Vol. 312, p. 1569, June 22, 1996.
- "Declining Blood Lead Levels and Changes in Cognitive Function during Childhood," Tong, Journal of the American Medical Association, Vol. 280, No. 22, 1998.
- "Do Low Levels of Lead Produce IQ Loss in Children?" Kaufman, Archives of Clinical Neuropsychology, Vol. 16, Issue 4, p. 303, May 2001.
- "Studies of Lead Exposure and the Developing Central Nervous System: a Reply to Kaufman," Needleman and Bellinger, Archives of Clinical Neuropsychology, Vol 16, Issue 4, p. 359, May 2001.

- "How Dangerous Are Low (Not Moderate or High) Doses of Lead for Children's Intellectual Development?," Kaufman, Archives of Clinical Neuropsychology, Vol. 16, Issue 4, p. 403, May 2001.
- "Childhood Blood Lead and Cognition," Gee and McKay, Clinical Toxicology, Vol. 40, No. 4, p. 519, 2002.
- "Intellectual Impairment in Children with Blood Lead Concentrations below 10 ug per Deciliter," Canfield et al., New England Journal of Medicine, Vol. 38, No. 16, p. 1517, April 17, 2003.
- Letter to the Editor, Bellinger and Needleman, The New England Journal of Medicine, Vol. 349, No. 5, July 31, 2003.
- Letter to the Editor, Canfield et al., The New England Journal of Medicine, Vol. 349, No. 5, p. 501, July 31, 2003.
- "Recent Developments in Low Level Lad Exposure and Intellectual Impairment in Children," Koller et al., Environmental Health Perspectives, vol. 112, No. 9, p. 987, June 2004.
- Letter to the Editor, Jusko et al., Environmental Health Perspectives, Vol. 113, No. 11, p. A16, Jan. 2005.
- Letter to the Editor, Koller et al., Environmental Health Perspectives, Vol. 113, No. 11. p. A16, Jan. 2005.
- "Changing Assumptions about Lead and IQ," Weinhold, Environmental Health Perspectives, Vol. 113, No. 5 p. A324, May 2005.
- "IQ and Blood Lead from 2 to 7 Years of Age," Chen et al., Environmental Health Perspectives, Vol. 113, No. 5, p. 579, May 2005.
- "Low-Level Environmental Lead Exposure and Children's Intellectual Function: An International Pooled Analysis," Environmental Health Perspectives, Vol. 113, No. 7, p. 894, July 2005.
- Letter to the Editor, Ernhart, Environmental Health Perspectives, Vol. 114, No. 2, p. A85, Feb. 2006.

- Letter to the Editor, Lanphear et al., Environmental Health Perspectives, Vol. 114, No. 2, p. A87, Feb. 2006.
- "Reduced Intellectual Development in Children with Prenatal Lead Exposure," Schnaas et al, Environmental Health Perspectives, Vol. 114, No. 5, p. 791 May 2006.
- "Lead Exposure in Children: Prevention, Detection and Management," American Academy of Pediatrics, PEDIATRICS, Vol. 116, No. 4, Oct. 2005.

V. THE CLEAN AIR ACT

A. General

The Clean Air Act (CAA) protects human health and the environment by regulating air quality and restricting emissions of pollutants from point sources to the air.[1] Among other things, the CAA directs the EPA to adopt regulations establishing national ambient air quality standards (NAAQS) and national emission standards for hazardous air pollutants (NESHAP). The CAA directs the EPA to adopt regulations governing motor vehicle fuels and fuel additives. NOTE: EPA air quality data show a 96% decrease in national average annual air emission concentrations during the years 1980 to 2005, reaching a level well under 0.5 ug/m^3.

The CAA applies to lead, lead alloys and compounds, and products containing lead in many ways. For example, lead compounds are listed as hazardous air pollutants in Section 7412(a), although Section 7412(b)(7) states that the EPA may not list elemental lead as a hazardous air pollutant. The EPA's regulations under the CAA apply to mining and smelting operations which produce lead and lead alloys. They apply to manufacturers using lead and its alloys and compounds to produce products such as batteries.

B. National Ambient Air Quality Standards (NAAQS)

In its NAAQS regulations, the EPA established an air quality standard for lead as particulate matter. The 24-hour average concentration limit is 150 ug/m^3, measured as PM 10. The annual arithmetic mean limit is 50 ug/m^3, measured as PM 10. 40 CFR § 50.6 and part 50, Appendix K.

In its NAAQS regulations, the EPA established national primary and secondary air quality standards for lead. The maximum arithmetic mean may not exceed 1.5 ug/cm^3, maximum arithmetic mean, averaged over a calendar quarter.[2]

The EPA's air quality standards are implemented by the state environmental agencies pursuant to state implementation plans (SIPs). The EPA establishes standards which the states must meet in their SIPs. Thus, the SIPs regulate lead emissions.[3]

---

[1] 42 U.S.C. § 7401 et seq.

[2] See 40 CFR § 50.12 and the Reference Method in Appendix G.

[3] 40 CFR Parts 51 and 52. See §51.117 on control strategies and numerous references in the state plans included in Part 52 and the quality surveillance provisions in Part 58.

Pursuant to the CAA, the EPA adopted regulations for the issuance of Primary Nonferrous Smelters Orders (NSOs) to lead and other primary nonferrous metal smelters unable to comply with SIP requirements for stack emissions of sulfur dioxide. An NSO deferred compliance based on a showing that there was no reasonably available control technology to meet the SIP.[4] The authority to grant NSOs expired January 1, 1988.

The EPA has been gathering and evaluating data and considering whether to revise its national air quality standard for lead. On July 31, 2006, the EPA requested comments on two parts of its document titled "Air Quality Criteria for Lead: Second External Review Draft." On September 29, 2006, the EPA published the final document "Air Quality Criteria for Lead" in response to a court order in a lawsuit by the Missouri Coalition for the Environment. Among other things, the document discussed the scientific data developed during the last two decades on lead in blood plasma and bone and on neurological effects on young children. The EPA indicated that the levels of exposure at which lead can have adverse health effects are lower than expected in documents which it prepared up to 1990. The EPA previously published a schedule for the preparation of an air quality criteria document, a staff paper, human health and ecological risk assessment reports, and a final promulgation notice by September 1, 2008.

C. New Source Performance Standards

The EPA has adopted new source performance standards (NSPS) for primary lead smelters[5] and secondary lead smelters.[6] The EPA has also adopted NSPS for lead-acid battery manufacturing plants[7] and metallic mineral processing plants.[8]

---

[4] 42 USC § 7419; 40 CFR Part 57.

[5] 40 CFR § 60.180 et seq.

[6] 40 CFR § 60.120 et seq.

[7] 40 CFR § 60.370 et seq.

[8] 40 CFR § 60.380 et seq.

(1) Primary Lead Smelters The EPA's new source performance standard (NSPS) for primary lead smelters sets limits on discharges to the atmosphere of particulate matter, sulfur dioxide and visible emissions (opacity). The standards apply to the sintering machine, sintering machine discharge end, blast furnace, dross reverberatory furnace, electric smelting furnace, and converter. The particulate matter standard limits discharges to 50 milligrams per dry standard cubic meter (50 mg/dscm). The sulfuric dioxide standard limits discharges 0.065% by volume. The standard for visible emissions is 20% opacity. The standard prescribes monitoring of operations, test methods and procedures.[9]

(2). Secondary Lead Smelters. The EPA's NSPS for secondary lead smelters limits the discharge of particulate matter into the atmosphere from a blast (cupola) or reverberatory furnace of gases which (1) contain particulate matter exceeding 50 mg/dscm or (2) exhibit 20% opacity or greater.[10]

(3) Lead-Acid Battery Manufacturers. The EPA's NSPS for lead-acid battery manufacturers sets limits for discharges to the atmosphere from any lead-acid battery manufacturing plant that produces or has the design capacity to produce in one day (24 hours) batteries containing 5.9 mg. (6.5 tons) of lead or more.[11] The standards apply to discharges of lead containing gases from the following facilities:

Grid casting	0.40 mg/dscm of exhaust
Paste mixing	1.00 mg/dscm of exhaust
Three process operation including plate stacking, burning or strap casting, and assembly into cases.	1.00 mg/dscm of exhaust
Lead oxide manufacturing	5.00 mg/dscm
Lead reclamation	4.50 mg/dscm
Other operations	1.00 mg/dscm of exhaust

---

9 40 CFR § 60.180 et seq.

10 40 CFR § 60.122.

11 40 CFR § 60.370 et seq.

The regulations require monitoring every 15 minutes according to prescribed test methods and procedures.

(4) Metallic Mineral Processors. The EPA's NSPs for metallic mineral processors sets limits for discharges into the atmospheres of any stack emissions that (1) contain particulate matter in excess of 0.05 grams per dscm, (2) exhibit greater than 7% opacity unless discharged from an affected facility using a wet scrubbing omission control device, and (3) with certain exceptions, discharge any process fugitive emissions that exhibit greater than 10% opacity.[12]

D. National Emission Standards for Hazardous Air Pollutants for Source Categories

(1) Secondary Lead Smelting

The CAA prohibits listing elemental lead as a hazardous air pollutant. However, the EPA is free to regulate lead compounds. Effective in 1995, the EPA adopted national emission standards for hazardous air pollutants (NESHAP) from secondary lead smelting. The NESHAP was amended 1997 and 1999.[13]

(a) MACT Requirements

The NESHAP applies to all secondary lead smelters, including both major sources and area sources. The NESHAP does not apply to primary lead smelters, lead refiners or lead remelters. Secondary lead smelters burning hazardous wastes solely for metal recovery continue to be exempt as provided in 40 CFR § 266.100(c). The standards require maximum achievable control technology (MACT) as required by the CAA. In general, the standards were based on control technology implemented by the five best-performing existing sources, subject to certain "floor" requirements.

(b) Limits on Emission Sources.

The standards impose limits on three sources. They are process emission sources, process fugitive sources and fugitive dust sources.

---

[12] 40 CFR § 60.380 et seq.

[13] 42 USC § 7412(b)(7); 40 CFR §63.541 et seq.

The process emission standards require that emissions of gases containing lead compounds from all types of secondary lead smelting furnaces must be limited to two milligrams of lead per dry standard cubic meter (2 mg/dscm). They also require that emissions of gases containing total hydrocarbons (THCs) be limited based on furnace types. Collocated blast and reverberatory furnaces must comply with a limit of parts per million by volume (20 ppm), expressed as propane, corrected to 4% carbon dioxide when both furnaces are operating. When the reverberatory furnace is not operating, blast furnaces for which construction or reconstruction commenced after June 9, 1994 must comply with a 70 ppm limit and earlier blast furnaces must comply with a 360 ppm limit. Smelters with blast furnace-only facilities for which construction or reconstruction commenced after June 9, 1994 must comply with a 70 ppm limit and earlier furnace-only facilities must comply with 360 ppm limit. If fugitive emissions from the charging chute are not combined with blast furnace emissions, the THC emission rate for the blast furnace process fugitive emissions must not exceed 0.2 kilograms per hour (0.2 kg/hr). Smelters which have only a reverberatory, rotary or electric furnace are not subject to a THC compliance limit.

The process fugitive emission standards require secondary lead smelters to comply with either of two control options. Each process fugitive emission source must be controlled by an enclosure hood ventilated to a control device or by a total containment building enclosure ventilated to a control device. Emissions from each control device must comply with a 2.0 mg/dscm limit. The standard prescribes minimum air (face) velocities for the enclosure hoods.

The fugitive dust emission standards require each secondary lead smelter to develop and submit to the EPA a standard operating procedure (SOP) manual. The SOP manual must describe in detail the measures that will be put in place to control fugitive dust emissions from the (1) plant roadways; (2) the battery breaking area; (3) the furnace area; (4) the refining and casting area; and (5) the materials storage and handling area. The controls specified in the SOP manual must include minimum requirements for each area. For example, the SOP manual must provide for paving of all plant roadways subject to vehicle traffic and, with some exceptions, for cleaning twice per day. Other minimum requirements include alternatives such as cleaning twice per day, a total enclosure and ventilation to a control device. Any gases containing lead compounds discharged from any ventilation system must not contain more than 2.0 milligrams of lead per day standard cubic meter (2.0 mg/dscm).

(c) Other Requirements of the NESHAP.

The NESHAP prescribes test methods and monitoring, notification, recordkeeping and reporting requirements. Among these numerous requirements, the monitoring requirements include preparation of an SOP manual describing in detail procedures for inspection, maintenance, and leak detection and corrective action plans for all baghouses (fabric filters) used to control emissions subject to the standard. They also require a continuous bag leak detection system using means such as triboelectric, light scattering or other effects to monitor relative or absolute particulate matter emissions. Recordkeeping and reports must evidence compliance with SOP manual procedures. Reports must include explanations of periods when procedures were not followed and corrective actions taken.

(2) Primary Lead Smelting

In its Fact Sheet dated July 1, 1999, the EPA said there are three primary lead smelters operating in the United States – ASARCO Inc. facilities at Glover, Missouri and East Helena, Montana and a Doe Run Company facility at Herculaneum, Missouri. Subsequently, ASARCO shut down its East Helena facility and filed for Chapter 11 bankruptcy proceedings. Thus, the NESHAP for primary lead smelting published by the EPA in the Federal Register on June 4, 1999 has limited application.[14]

The primary lead smelting NESHAP sets a "plant wide" emission limit of 500 grams of lead per megagram of lead metal produced (1.0 pounds of lead per ton of lead metal produced) from the aggregation of emissions discharged from the air pollution control devices from the following sources: (1) sinter machine; (2) blast furnace; (3) dross furnace; (4) dross furnace charging location; (5) blast furnace and dross furnace tapping location; (6) sinter machine charging location; (7) sinter machine discharge end; (8) sinter crushing and sizing equipment; and (9) sinter machine area. Standards for fugitive dust sources are also prescribed.

[14] 64 Fed. Reg. 30204, June 4, 1999; 40 CFR §63. 1541 et seq.

The NESHAP also prescribes test methods and monitoring, notification, recordkeeping and reporting requirements.

E. Regulation of Fuel and Fuel Additives

The CAA authorizes the EPA to control or prohibit the manufacture or sale of any motor vehicle fuel or fuel additive which will reasonably be anticipated to endanger the public health and welfare.[15] The EPA adopted regulations mandating the manufacture and sale of unleaded gasoline, defined as gasoline containing no more than 0.05 gram of lead per gallon and not more than 0.005 gram of phosphorus per gallon.[16] The EPA developed the regulations into a phaseout of leaded gasoline, i.e., any gasoline containing more than 0.05 grams of lead per gallon or more than 0.005 grams of phosphorous per gallon.[17] Sale of leaded gasoline for motor vehicle use was prohibited after December 31, 1995.[18]

---

[15] 42 USC § 7545(c).

[16] 40 CFR § 80.2(g).

[17] 40 CFR § 80.22 et seq.

[18] 42 USC § 7545(n).

VI. THE CLEAN WATER ACT

A. General

The Clean Water Act (CWA) was adopted to protect the waters of the United States from discharges of hazardous substances and toxic and other pollutants.[1] Among other things, the CWA requires that spills of hazardous substances in excess of threshold quantities be reported to the EPA. The CAA establishes the national pollutant discharge elimination system (NPDES) under which industrial, municipal and other dischargers of effluents containing pollutants must obtain and comply with permits regulating and limiting such discharges. The CWA directs the EPA to establish water quality standards, effluent guidelines and standards, and other regulations including guidelines for the management of sewage sludge. The EPA has adopted extensive regulations implementing the CWA.[2]

B. Listing of Lead Compounds as Hazardous Substances

Several lead compounds are included in the EPA's list of hazardous substances required by the CWA.[3] They are lead acetate, lead arsenate, lead chloride, lead fluoborate, lead fluoride, lead iodide, lead nitrate, lead stearate, lead sulfate, lead sulfide and lead thiocyanate.

C. Reportable Quantities for Spills of Lead Compounds

The EPA has listed reportable quantities for spills of hazardous substances.[4] The Table establishes the following reportable quantities for spills of lead compounds:

Lead arsenate - 1 pound

Other lead compounds - 10 pounds each

D. National Pollutant Discharge Elimination System

The EPA has adopted extensive regulations to implement the NPDES systems.[5] Lead is listed as a toxic pollutant.[6] Discharges containing lead from point sources are subject to

---

[1] 33 USC § 1251 et seq.

[2] CFR Parts 100 to 149 and Parts 401 to 503.

[3] 40 CFR § 116.4, Table 116.4.

[4] 40 CFR § 117.3, Table 117.3.

[5] 40 CFR Part 122.

[6] Part 122, et seq. Appendix D, Table III.

the permit program established by the NPDES regulations including the testing requirements.[7] The point sources include not only industrial and commercial discharges, but also those from publicly owned treatment works, stormwater, and certain agricultural and silvicultural operations.[8]

The EPA has adopted regulations providing criteria and standards for the NPDES system.[9] The criteria are technology-based and permits must require best available technology (BAT) for toxic pollutants including lead.[10] The regulations also contain criteria for special permits, extension, variances, modified secondary treatment requirements, best management practices, and sewage sludge disposal. Some of these provisions define toxic pollutants by reference to the list in the EPA's Effluent Guidelines and Standards which include lead and its compounds, both organic and inorganic.[11]

The NPDES regulations prescribe test procedures for measurement of lead for the purpose of permit applications, reports and certifications under the NPDES system.[12] They require digestion followed by atomic absorption using direct aspiration or furnace equipment.

E. State NPDES Program Requirements

Each state administering its own NPDES program was required to submit its program and obtain approval by the EPA.[13] To obtain approval, the program must have established permit requirements at least as stringent as the EPA's requirements.[14] Thus, state

---

[7] 40 CFR § 122.21.

[8] 40 CFR § 122.1.

[9] 40 CFR Part 125.

[10] 40 CFR § 125.3.

[11] See 40 CFR § 401.15.

[12] 40 CFR § 136.3, Table 1B(32).

[13] 40 CFR Part 123.

[14] 40 CFR § 123.25.

environmental agencies must regulate discharges of lead and lead compounds as strictly as the EPA.

F. Water Quality Management and Planning Regulations

The EPA has adopted water quality management and planning regulations at 40 CFR, Part 130 that include lead and lead compounds.

G. Water Quality Standards for State Programs

The EPA has adopted regulations governing state water quality standards.[15] They require state standards to meet minimum requirements, including an antidegradation policy, and to be reviewed and approved by the EPA. Lead is among the priority toxic pollutants for which freshwater and saltwater standards must be established. For states not complying with its procedures, the EPA has set numerical criteria for priority toxic pollutants, including lead, in 40 CFR §131.36. For freshwater, the maximum concentration limit is 65 ug/L and the continuous concentration limit is 2.5 ug/L. For saltwater, the maximum concentration limit is 210 ug/L and the continuous concentration limit is 8.1 ug/L. No human health standard is prescribed.

H. Toxic Pollutant Effluent Standards

The EPA's toxic pollutant effluent standards in 40 CFR Part 129 do not list lead or lead compounds among the toxic pollutants which are subject to its requirements and which may be incorporated into any NPDES permit, modification or renewal.

I. Effluent Guidelines and Standards for Classes and Categories of Sources

The EPA has set effluent guidelines and standards for specific classes and categories of point sources.[16] They prescribe effluent limitations for existing sources, standards of performance for new sources, and pretreatment standards for new and existing sources.[17]

---

[15] 40 CFR Part 131.

[16] 40 CFR Parts 401 to 471.

[17] 40 CFR § 401.10.

They contain a list of toxic pollutants which includes lead and its compounds, both organic and inorganic.[18]

<u>Mining and Milling of Ores</u>. The EPA has set effluent limitations for mines that produce copper, lead, zinc, gold, silver, or molybdenum ores (or any combination of them) from open pit or underground operations other than placer deposits. The regulations also set limitations for mills that use the froth flotation process alone or with other processes for the bonification of copper, lead, zinc, gold, silver, or molybdenum ores, or any combination of them. The limits for existing point at mines and mills required to comply with the best practicable control technology currently available (BPT) are (1) a maximum per day of 0.6 mg/L and (2) an average of daily values for 30 consecutive days of 0.3 mg/L. These limits apply both to mine drainage and mill discharges.[19] The same limits apply to mines and mills required to comply with best available technology economically achievable (BATEA).[20]

<u>Primary Lead Smelting and Refining</u>. The EPA has set effluent limitations for primary lead smelters and refiners.[21]

The regulations begin with limitations based on best practicable control technology currently available (BPT) measured in pounds per million pounds for one day and for a monthly average for the (a) sinter plant wet air pollution control, (b) blast furnace wet air pollution control, (c) blast furnace slag granulation, (d) dross reverberatory slag granulation, (e) dross reverberatory furnace wet air pollution control, (f) zinc fuming wet air pollution control, (g) hard lead refining slag granulation, (h) hard lead refining wet air pollution control, (i) facility washdown, (j) employee handwash, (k) respirator wash, and (l) uniform laundering. No emissions are allowed from the blast furnace wet air pollution control, the zinc fuming wet air pollution control and the facility washdown sources.

---

[18] 40 CFR, § 401.15.

[19] 40 CFR § 440.102.

[20] 40 CFR § 440.103.

[21] 40 CFR § 421.70.

The regulations establish strict standards based on best available technology economically achievable (BATEA). For example, the BATEA does not allow any emissions from the blast furnace wet air pollution control or slag granulation, the dross reverberatary furnace wet air pollution control, the zinc fuming wet air pollution control, the hard lead slag granulation or wet air pollution control, or the facility washdown.

The regulations then establish still stricter new source performance standards (NSPS). The NSPSs prohibit omissions from the same sources as the BATEA regulations and also prohibit omissions from the sinter plant materials handling wet air pollution control and the dross reverberatory slag granulation.

The regulations establish pretreatment standards for primary lead smelters discharging wastewater to a publicly owned treatment works (POTW). The pretreatment standards for existing sources (PSES) correspond to the BATEA limitations. The pretreatment standards for new sources (PSNS) correspond to the NSPS limitations.

Secondary Lead Smelting and Refining. The EPA has set effluent limitations for secondary lead facilities.[22] The regulations begin with limitations based on BPT. They prescribe numerical maximum daily and monthly average limits in pounds per million pounds for battery cracking; blast reverberatory or rotary furnace air pollution control; kettle wet air pollution control; lead paste desulfurization; casting contact cooling; facility washdown; battery case classification; employee handwash; employee respirator wash; and laundering of uniforms. The regulations do not allow any emissions from the lead paste desulfurization, the facility washdown and the battery case classification sources.

The regulations then establish stricter BATEA effluent limitations and NSPS limitations which are the same except that the NSPS limitations do not allow any emissions from the kettle wet air pollution control. The regulations also establish pretreatment standards for secondary smelters discharging wastewater to a POTW. The PSES limitations correspond to the BATEA limitations. The PSNS limitations correspond to the NSPS limitations.

---

[22] 40 CFR § 421.130.

Battery Manufacturing. The EPA has set effluent limitations for battery manufacturing plants.[23]

The standards are divided into subcategories based on battery anodes: cadmium, calcium, lead, Le Clanche (zinc), lithium, magnesium and zinc. For the cadmium, lead and zinc subcategories, the regulations set BPT, BATEA, NSPS, PSES and PSNS limitations. For the Le Clanche and magnesium subcategories, the regulations set only NSPS, PSES and PSNS limitations. For the calcium and lithium subcategories, the regulations set only NSPS and PSNS limitations.

Other Source Categories

The regulations also contain effluent limitations for discharges of lead from point sources at numerous other source categories. The standards include technology-based effluent limitations, i.e., BPT, BATEA, NSPS, PSES and PSNS. The categories include electroplating (Part 413), inorganic chemicals manufacturing (Part 415), iron and steel manufacturing (Part 420), nonferrous metals manufacturing (Part 421), steam electric power generating (Part 423), metal finishing (Part 433), metal molding and casting (Part 464), porcelain enamel (Part 466), copper forming (Part 468), electrical and electronic components (Part 469), and nonferrous metals forming and metal powders (Part 471).

Some of the categories include numerous subcategories in which effluent limitations for lead are set. For example, among the subcategories for which lead limitations are set in the nonferrous metals manufacturing subcategory are secondary aluminum smelting, primary electrolytic copper, primary zinc, metallurgical acid, primary tungsten, primary tantalum, primary and secondary germanium, secondary indium, secondary mercury, primary molybdenum and rhenium, secondary molybdenum and vanadium, primary precious metals and mercury, primary rare earth metals, secondary tin, primary and secondary titanium, and primary zirconium and hafnium.[24]

---

[23] 40 CFR Part 461.

[24] See generally 40 CFR Parts 401-471.

J. Sewage Sludge Management

The EPA's sewage management regulations restrict the concentrations and loading rates of lead in sewage sludge applied to land. The ceiling concentration 840 mg/kg.[25] The cumulative pollutant loading rate limit is 300 kg per hectare.[26] The monthly average concentration limit is 300 mg/kg.[27] The annual pollutant loading rate limit is 15 kg per hectare per 365 day period.[28]

[25] 40 CFR § 503.12, Table 1.

[26] 40 CFR § 503.13, Table 2.

[27] 40 CFR § 503.13, Table 3.

[28] 40 CFR § 503.13, Table 4.

VII. THE SAFE DRINKING WATER ACT (SDWA)

A. General

The Safe Drinking Water Act (SDWA) was adopted to protect the nation's drinking water.[1] The SDWA requires the EPA to set standards for contaminants in drinking water. The drinking water standards apply to public water systems which provide piped water for human consumption having at least 15 service connections or regularly serving at least 25 individuals, but excluding hotels and other storage and distribution facilities which obtain water from a public water system. The states have primary enforcement responsibility for the SDWA, provided they adopt regulations as strict as those of the EPA and meet other requirements set by the EPA.

B. Prohibited Uses

The SDWA prohibits the use of water pipes, fittings, fixtures, solder or flux that is not "lead free," i.e., containing more than 8% lead and solder and flux containing more than 0.2% lead, in any public water system or any plumbing in a residential or nonresidential facility providing water for human consumption, but allows an exception for leaded joints necessary for cast iron pipes.[2] The EPA's regulations repeat the statutory prohibitions and exceptions.[3]

The SDWA requires public water systems to provide notice to persons who may be affected by lead in their drinking water resulting from lead content in the construction materials of their distribution systems or corrosivity of the water supply sufficient to cause leaching of lead. An example of the notice appears as Appendix A to this Chapter.

The SDWA prohibits any person (A) to introduce into commerce any pipe, fitting or fixture that is not lead free except for a pipe used in manufacturing or industrial processing, (B) engaged in the business of selling plumbing supplies, except manufacturers, to sell solder or flux that is not lead free, and (C) to introduce into commerce any solder or flux that is not lead

---

[1] 40 USC § 300 <u>et</u> <u>seq.</u>

[2] 42 U.S.C. § 300g-6.

[3] 40 CFR § 141.43.

free without label warning against use in installation or repair of plumbing providing water for human consumption.

The SDWA also prohibits and recalls drinking water coolers containing any component which may contact drinking water containing more than 8% lead or any solder, flux or interior surface which may contact water containing more than 0.2% lead.[4] A requirement in Section 300 j-24(d) that states establish remedial action programs relating to drinking water and other sources of lead contamination in schools was held unconstitutional.[5]

C. Primary and Secondary Drinking Water Standards

The EPA has adopted primary and secondary drinking water standards. The primary standards are health-based standards which apply to contaminants found in drinking water that may have an adverse effect on human health. The secondary standards are not health-based and are issued as guidelines to state agencies in relation to substances which may adversely affect the odor or appearance of water or may cause a substantial number of persons to discontinue use. They are not enforceable by the USEPA against public water systems.

The SDWA contains strict procedures for determining standards. The procedures begin with setting maximum contaminant level goals (MCLGs) which are nonenforceable. They are based on reference doses derived from no observed effect levels (NOAELs) which are derived from studies which are often quite limited and suggest adverse effects based on modest evidence. The MCLGs tend to be set at extremely low levels. For substances which environmental researchers suspect to be carcinogens, the MCLG is set at zero. The EPA then sets the maximum containment level (MCL) as close to the MCLG as can be achieved by the best available technology, taking into account cost and effectiveness in actual operation. The SDWA allows public water systems to apply for variances and exemptions from compliance with the MCLs, but they are for short periods and are not easily obtained.

---

[4] 42 U.S.C. § 300 j-21 to24 including provisions of the Lead Contamination Control Act.

[5] ACORN v. Edwards, 81 F.3d 1387 (5th Cir. 1996); cert. den. 521 U.S. 1129 (1997).

D. National Primary Drinking Water Regulations for Lead

The EPA's national primary drinking water regulations set the maximum containment level goal (MCLG) for lead at zero.[6] The regulations formerly established a maximum containment level (MCL) of 0.05 milligrams per liter (mg/L.)[7] However, the MCL expired on November 9, 1992 and was replaced by action level requirements and other requirements imposed on community water systems.[8] Community water systems are required to install and operate corrosion control treatment to minimize lead and copper concentrations.[9] They must also meet monitoring, analytical, recordkeeping and reporting requirements. Any community water system exceeding the action level of 0.015 mg/L (also expressed as 15 parts per billion, or 15 ppb) for 10% of its tap water samples must also meet source water treatment, lead service line replacement and public education requirements.[10] For example, community water systems failing to meet the action level must replace seven per year of the lead service lines they control until the action level is met for two consecutive monitoring periods.

In 1993, The American Water Works Association (AWA), an industry association, and the Natural Resources Defense Council (NRDC), an environmental organization, challenged the EPA's new regulations in the U.S. Circuit Court of Appeals for the District of Columbia. The Court upheld the AWA's argument that the EPA's definition of control in the regulations excessively extended the responsibilities of community water systems to include facilities they do not own. The Court rejected the NRDC's argument that the regulations should be declared invalid because the EPA was required to set an MCL for lead.[11]

---

[6] 40 CFR § 141.51.

[7] 40 CFR § 141.51.

[8] 40 CFR § 141.80.

[9] 40 CFR § 141.82.

[10] 40 CFR § § 141.83 to 85.

[11] American Water Works Association v. Environmental Protection Agency, 40 F.3d 1266 (D.C. Cir. 1994).

On July 18, 2006, the EPA announced proposed changes to the lead and copper regulations to add to the public education, corrosion control treatment, source water treatment, lead service line replacement, and supplemental monitoring requirements.[12]

E. National Secondary Drinking Water Regulations for Lead

The EPA has adopted national secondary drinking water regulations.[13] They apply to contaminants that primarily affect the aesthetic qualities relating to public acceptance of drinking water. No secondary MCL has been established for lead or lead compounds.

---

[12] 71 Fed. Reg. 40, 828 (July 18, 2006).

[13] 40 CFR Part 143.

Can the Risk be Lowered?

To reduce the amount of lead in water:

- Run the tap until water is cold to the touch before using it for drinking or cooking. (The flushed water can be saved for watering household plants, washing dishes or general household cleaning.)
- Check household plumbing for lead-based pipes or solder.
- Use only lead-free materials in all plumbing repairs or installations. The use of lead solder in plumbing was banned in New York State in 1986. Ask the plumber to show you the label from any solder packaging being used. It should state that the solder is lead-free.
- Use only water from the cold water tap for cooking, drinking, or making a baby's formula. Hot water promotes leaching.

How can I test for Lead?

Certified commercial laboratories can test for lead in drinking water. The cost ranges from $15 to $50 per sample. Contact the Westchester County Department of Health or the New York State Department of Health for the names of approved local laboratories.

Consumer Beware: Unscrupulous businessmen will use tests or offer filtering devises that have not been found to be effective. Use only professional laboratories.

What is the Level of Lead in Public Drinking Water Supplies?

In July 1991, the United States Environmental Protection Agency (EPA) established an action level for lead in drinking water at 15 ug/l (micrograms per liter), or 15 parts per billion (ppb). Water suppliers are required to monitor household tap water and , depending on the size of the population served, must follow a prescribed schedule to initiate methods to reduce corrosion and lower lead levels of household tap water.

In compliance with the Safe Drinking Water Act, the City of Yonkers Water Bureau has sampled drinking water from homes in our distribution system for the presence of lead and copper. The results indicate that SOME HOMES IN THIS COMMUNITY HAVE ELEVATED LEAD LEVELS IN THEIR DRINKING WATER above 15 micrograms per liter and/or copper levels above 1.3 milligrams per liter. These are levels which the United States Environmental Protection Agency has set to determine when a water supplier must take action to reduce the amount of lead and/or copper in the drinking water.

Lead and copper concentrations increase in drinking water when the water comes in contact with lead and copper piping, lead solder joints and brass faucets and fittings. The longer water is in contact with these materials, the higher the concentration of lead and cooper in the drinking water. These materials, lead, copper and brass, are **primarily present in the home plumbing system**. You can avoid drinking water with lead or copper by flushing the line for a few minutes before using the water for drinking. LEAD CAN POSE A SIGNIFICANT RICK TO YOUR HEALTH. PLEASE READ THIS BROCHURE FOR FURTHER INFORMATION.

YONKERS WATER BUREAU
1070 Nepperhan Avenue
Yonkers, N.Y. 10703
(914) 377-6006

State Approved Lead Testing Laboratories:
Westchester County Laboratory (914) 593-5575
Consolidated Technology (914) 591-9010
Blood Testing: Westchester County Health Dept. (914) 593-5100

FLUSHING TAP TWICE A DAY COSTS LESS THAN 25 CENTS PER MONTH

State of New York
George E. Pataki, Governor

Department of Health
Mark R. Chassin, M.D. , Commissioner

City of Yonkers
John D. Spencer, Mayor

U.S. Postage
PAID
Permit No. 256
Yonkers, NY

GET THE LEAD OUT OF DRINKING WATER

What is Lead

Lead is a metal found naturally in soil and in both ground and surface waters. People may be exposed to lead in the food they eat, the air they breathe, and the water they drink. But lead can be harmful to health and can cause problems when it accumulates in the body. The danger from exposure depends on the amount of lead, the length of time a person is exposed to lead, and the person's age.

Why the Concern?

Too much lead in the human body can cause serious damage to the brain, nervous system, and red blood cells. Pregnant woman and young children are at the greatest risk even with short-term, low level exposures. In fact, young children are more susceptible than other people to lead, and more likely to suffer health problems. Lead poisoning can slow a child's normal growth, and physical and cognitive development; and can cause behavior problems mental retardation, kidney and liver damage, blindness, and even death.

How Do I Know If My Family Has High Blood Lead Levels?

If the drinking water in your home has been tested for lead, and the results confirm lead levels higher than 20 ug/l, see your doctor and tell him or her that your family has been drinking water that contains high lead levels. Pregnant women should

be sure to mention this possibility to their obstetricians.

If you have questions concerning testing for lead in your drinking water or if you want advice on how to lower the lead levels in your drinking water, contact the Westchester County Health Department at 593-5100.

The New York State Department of Health also offers information. For medical concerns, call the Bureau of Environmental and Occupational Epidemiology, toll-free, at 1-800-458-1158.

For more information on water testing or interpretation of results, call the Bureau of Public Water Supply Protection at 518-458-6731

How Does Lead Get into the Water We Drink?

Since natural levels of lead in New York State water supplies are low, elevated levels of lead usually result from the use of lead pipe in water systems or lead-based solder on water pipes.

Lead in drinking water may occur when there is leaching of the lead from pipe or solder. Leaching occurs when the action of corrosive water removes mineral deposits from the inside of water pipes exposing the lead pipe or solder to the water. Soft (corrosive) or acidic (low pH) water is more likely to cause leaching.

Water left stagnant in the pipes over a long period of time also results in leaching. The longer the water stands in the pipes, the greater the possibility of lead being dissolved into the water, Stray electrical currents from improperly grounded electrical outlets or equipment also may increase the level of lead in drinking water. And the distribution system that carries drinking water from the source to residences can contribute to the drinking water, if lead-based piping materials were used during construction or repair.

Lead levels in drinking water are more likely to be high if:

- the home or water distribution system had lead pipes
- the home has copper pipes with lead solder
- the home is less than five years old (or until the pipe is coated with mineral deposits and/or lead is leached from the solder)
- the water is either soft (corrosive) or acidic (low pH)
- the water sits unused in pipes for long pipes of time
- the water pipes in the home are used as an electrical ground

VIII. THE EMERGENCY PLANNING AND COMMUNITY RIGHT-TO-KNOW ACT

A. General

The Emergency Planning and Community Right-To-Know Act (EPCRA) was adopted in 1986 to help states and communities to plan for and respond to potential emergencies involving hazardous substances.[1]

B. Emergency Planning and Release Reporting

EPCRA required that the EPA publish a list of extremely hazardous substances and provides that any facility having more than a threshold quantity of any listed substance must comply with emergency planning and release reporting requirements. The list published by the EPA does not include lead but includes the compounds, tetraethyl lead and tetramethyl lead.[2] The reportable quantity for a release of tetraethyl lead is 10 pounds. The threshold planning quantity is 100 pounds. The reportable quantity and threshold planning quantity for tetramethyl lead are both 100 pounds. Because of the phaseout of tetraethyl lead as a gasoline additive mandated by the Clean Air Act, these requirements are no longer as important as in prior years.

C. Communication of Chemical Hazards to Community Officials

EPCRA also requires the owner or operator of each facility where hazardous chemicals or extremely hazardous substances are present in excess of certain threshold quantities to furnish a material safety data sheet (MSDS) for each hazardous chemical to the state emergency response commission (SERC), the local emergency response committee (LERC) and the fire department. The owner or operator must also furnish to them an annual inventory of hazardous chemicals present at the facility above specified threshold levels.[3] For the purpose of these regulations, a hazardous chemical is defined by reference to OSHA's hazard communication regulations.[4] Lead is considered to be a hazardous chemical under those

---

[1] 42 USC § 11001 et seq.

[2] 40 CFR Part 355, Appendix A.

[3] 40 CFR Part 370.

[4] 29 CFR § 1910.1200.

regulations.[5] Thus, owners and operators of facilities where lead is present in excess of threshold quantities must furnish an MSDS and the annual inventory reports to the SERC, LERC and fire department. Forms for reporting Tier I and also for Tier II information, if needed, are prescribed by the EPA's regulations.[6]

D. Toxic Chemical Release Reporting

EPCRA also requires facilities in Standard Industrial Codes 20 through 39 with ten or more full-time employees to report to the EPA and state environmental agencies their releases of toxic chemicals that can be reasonably expected to cause adverse human health effects or significant adverse effects on the environment. The reports must be made on a Toxic Chemical Release Inventory Reporting Form (Form R). Lead and lead compounds are included in the list of toxic chemicals subject to these reporting requirements.[7]

The EPCRA regulations class lead and lead compounds among the persistent, bioaccumulative and toxic (PBT) chemicals for which specially low reporting thresholds are set. Facilities which manufacture, process or use 100 pounds or more of lead and lead compounds are subject to toxic release inventory (TRI) reporting without availability of the de minimis exemption and the million pound alternate threshold. 40 CFR §372.28. See 66 Fed Reg. 4499 (Jan. 17, 2001) and 64 Fed. Reg. 58,666 (Oct. 29, 1999).

[5] See 29 CFR § 1910.1200(d)(3)(i) and 29 CFR § 1910.1000, Table Z-1.

[6] 40 CFR § 370.40 and .41.

[7] 40 CFR § 372.65(a) which lists lead and (c) which lists lead compounds.

IX. THE SOLID WASTE DISPOSAL ACT, AS AMENDED BY THE RESOURCE CONSERVATION AND RECOVERY ACT

A. General

The Solid Waste Disposal Act (SWDA) was a relatively limited law until it was amended extensively in 1976 by the Resource Conservation and Recovery Act (RCRA). As a result, the law is commonly called "RCRA."[1]

RCRA provides that the states have primary responsibility for management of nonhazardous solid waste, but directs the EPA to adopt regulations setting minimum standards which the states must meet.

RCRA also directs the EPA to establish, administer and enforce regulatory systems governing the management of solid waste including hazardous waste. The EPA's first major program under RCRA was the adoption and implementation of its Hazardous Waste Management System (HWMS) regulations governing hazardous waste from generation through disposal and post-disposal care, often called a "cradle to the grave" system.[2] Key elements of the HWMS require the use of a Uniform Hazardous Waste Manifest Form 8700-22 prescribed by the EPA for every transportation of hazardous waste for offsite treatment, storage or disposal and compliance with pre-transport, recordkeeping and reporting requirements.

As described, RCRA is a waste management law. RCRA requires cleanup and other corrective actions at facilities subject to its jurisdiction such as permitted treatment, storage and disposal (TSD) facilities. However, RCRA is not primarily intended to require the cleanup of hazardous substances and wastes resulting from disposal activities prior to its enactment. That task is governed by the Comprehensive Environmental Response, Compensation and Liability Act and state laws described in Chapter X. If a TSD facility was properly closed and no longer in existence before November 19, 1980, a TSD permit is not required.

---

[1] 42 USC § 6901 et seq.

[2] 40 CFR Part 262.

B. Solid Waste Disposal Facilities

The EPA has established criteria for classification of solid waste disposal facilities and practices. The regulations include maximum contaminant levels (MCLs) for groundwater beyond the solid waste boundary (or an alternative boundary). The MCL for lead continues to be listed as 0.05 mg/L, although the MCL under the SDWA has been changed to an action level.[3]

The EPA has also established criteria for municipal waste landfills (MSWLF units) which their owners and operators must meet so that the facilities will not be deemed "open dumps" prohibited by RCRA. The design criteria for MSWLF units must ensure that lead concentration in the uppermost aquifer will not exceed the MCL of 0.05 mg/L.[4]

C. The Hazardous Waste Management System Regulations

Under RCRA, a substance must first be a solid waste before it can be a hazardous waste. This threshold issue has significance for many materials containing substances, such as asbestos or lead, that are classed as hazardous under other laws. For example, properly maintained asbestos-containing insulation and lead-based paint on building walls and fixtures is not solid waste. Thus, although the insulation or paint contains a hazardous material, it is not hazardous waste until it becomes a discarded material which is abandoned, recycled or inherently waste-like.[5]

A solid waste material is a hazardous material under RCRA if (1) it exhibits characteristics of hazardous waste, (2) is listed by the EPA as hazardous waste or (3) is a mixture of solid and hazardous wastes, and (4) is not excluded from regulation as a hazardous waste.[6] The EPA's characteristics of ignitability, corrosivity and reactivity do not apply to lead and most

---

[3] 40 CFR § 257.3-4 and 40 CFR Part 257, Appendix I.

[4] 40 CFR § 258. 40, Table 1.

[5] 40 CFR § 261.2.

[6] 40 CFR § 261.3.

of its alloys and compounds. However, the toxicity characteristic applies to any solid waste from which the extract from a representative sample equals or exceeds a concentration of 5.0 mg/L using a Toxic Characteristic Leaching Procedure (TCLP) Test Method.[7]

Lead itself is not listed as hazardous waste. However, several solid wastes have been listed by the EPA as hazardous wastes because of their lead content.[8] Lead is listed as toxicity characteristic waste D008 at 5.0 mg/L, or 100 times the primary drinking water MCL. In the lists of hazardous wastes, the "F" series hazardous wastes from nonspecific sources in §261.31 do not include any wastes listed because of lead or its compounds except F035 wood preserving wastes. The "K" series hazardous wastes from specific sources in §261.32 include several wastes listed partially because of their lead content:

K002	Wastewater treatment sludge from the production of chrome yellow and orange pigments.
K003	Wastewater treatment sludge from the production of zinc yellow pigments.
K005	Wastewater treatment sludge from the production of chrome green pigments.
K006	Wastewater treatment sludge from the production of chrome oxide green pigments (anhydrous and hydrated).
K007	Wastewater treatment sludge from the production of iron blue pigments.
K008	Oven residue from the production of chrome oxide green pigments.
K048	Dissolved air flotation (DAF) float from the petroleum refining industry.
K049	Slop oil emulsion solids from the petroleum refining industry.
K051	API separator sludge from the petroleum refining industry.
K061	Emission control dust/sludge from the primary production of steel in electric furnaces.

[7] 40 CFR § 261.24, Table 1.

[8] 40 CFR §§ 261.30 to 261.33.

K069	Emission control dust/sludge from secondary lead smelting. (NOTE: This listing is stayed administratively for sludge generated from secondary acid scrubber systems. The stay will remain in effect until further administrative action is taken. If EPA takes further action effecting this stay, EPA will publish a notice of the action in the Federal Register.
K100	Waste leaching solution from acid leaching of emission control dust/sludge from secondary lead smelting.

The "P" series of discarded commercial chemical products, off-specification species, container residues, and spill residues in §261.33 includes only tetraethyl lead, P110.

The "U" series of commercial chemical products, manufacturing chemical intermediates, or off-specification commercial chemical products in §261.33 include some hazardous wastes listed because of their lead content; U144 (lead acetate), U146 (lead, bis) (acetato-0) tetrahydroxy -, U145 (lead phospate), and U146 (lead subacetate).

Sampling and analysis methods for lead are prescribed in the regulations.[9] Flame and furnace atomic absorption spectrometric methods are specified.

D. Land Disposal Restrictions

The EPA's land disposal regulations under RCRA, sometimes called the "Land Ban Regulations" impose restrictions on land disposal of certain hazardous wastes unless they have been treated to meet prescribed treatment standards.[10]

Wastes containing lead or lead compounds are restricted wastes if they exceed 5.0 mg/L as measured by the TCLP test or are one of the "F", "K", "P" or "U" listed wastes.[11] Generators must analyze their wastes to determine whether they have a toxic characteristic by using the TCLP test.[12]

---

[9] 40 CFR §260.11 and SW-846, Third Edition and updates.

[10] 40 CFR Part 268.

[11] 40 CFR Part 261, Appendix VII.

[12] 40 CFR § 268.7.

The EPA's regulations prescribe the treatment standards which must be met to allow disposal of a restricted waste including the attached tables which list lead as the regulated hazardous constituent of many of the restricted wastes. [13]

E. Management of Spent Lead-Acid Batteries

Partial exemptions from the hazardous waste management requirements are provided for lead-acid batteries being reclaimed to the extent allowed by 40 CFR §266.80.

F. Universal Waste Management Standards

Certain spent lead-acid batteries not reclaimed under 40 CFR §266.80 are subject to the universal waste management standards at 40 CFR Part 273.

G. Disposal of Lead-Based Paint Wastes

The requirement to comply with the EPA's regulations under RCRA is an important problem when lead-based paint is removed during abatement. They may be toxic characteristic hazardous wastes if they equal or exceed 5.0 mg/L when tested using the TCLP method. The time and cost of TCLP tests add significantly to the difficulty and cost of lead-based paint abatement.

In March, 1993, the EPA published the results of a study which its Office of Pollution and Toxics performed to provide assistance to abatement contractors experiencing difficulty in finding practical ways to dispose of waste in compliance with RCRA. In the Executive Summary, the EPA described the study, in part, as follows:

> "First, data on waste testing from the Department of Housing and Urban Development's (HUD's) nationwide abatement demonstration project was evaluated to determine the hazardousness as defined under RCRA of various categories of abatement waste. Second, EPA designed and conducted a detailed testing program for two important categories of waste not adequately tested in the HUD demonstration (large solid debris, and

[13] 40 CFR § 268 40 et seq.

protective plastic sheeting). Third, waste disposal experience of HUD's contractor for the demonstration project was examined to obtain preliminary estimates of the volume of hazardous waste generated, and the cost of disposing of these wastes.

There are three categories of waste produced in large volume during lead-based paint abatement. These are:

1. Filtered wash-water;
2. Solid debris, such as old woodwork, plaster, windows, doors, and similar bulky components;
3. Plastic sheets and tape used to cover floors and other surfaces.

Filtered wash-water was found to be non-hazardous, and may be disposed of according to State and local requirements. Solid debris was generally found to be hazardous only if the lead level in the paint, as measured in the laboratory by Atomic Absorption Spectrometry (AAS), exceeded approximately 4.0 milligrams per square centimeter (mg/cm^2). Specifically, 5 of 6 samples tested (83%) whose paint AAS lead level exceeded 4.0 mg/cm^2 failed the Toxicity Characteristic Leaching Procedure (TCLP) test for lead toxicity. Conversely, only 1 of 14 (6%) with paint AAS lead level below 4.0 mg/cm^2 failed the TCLP test.

Measurements of paint lead level by field X-Ray Fluorescence (XRF) were poorly correlated with TCLP results. Due to the limited number and non-random selection of samples tested, the results for solid debris are suggestive only, and require confirmation in a larger study before they can be used as a basis for EPA policy.

Plastic sheeting was found to be hazardous if certain abatement methods were used. Based on the samples available to us, when a heat gun was used for paint removal, the plastic sheeting was hazardous. However, it should be noted that, because of the limited availability of samples, all plastic sheeting samples tested came from a single dwelling; it would be desirable to confirm the above finding by testing additional plastic samples from other dwellings where a heat gun was used. When encapsulation (use of flexible wall covering systems of a reinforced fiber type that form a secure bond with the substrate) or enclosure (covering LBP-contaminated surfaces with wood paneling, gypsum board, or fabricated exterior enclosure systems of aluminum, vinyl or wood) were used, the plastic was sometimes marginally hazardous. When other methods were used, in particular removal and replacement of contaminated components, the plastic was non-hazardous. It should be noted that, although the above findings are indicative of whether or not various types of abatement waste are likely to be hazardous, waste generators are ultimately responsible under EPA regulations for the proper characterization and disposal of their waste.

There are several other categories of waste commonly produced during abatement, such as sludges, paint chips, mops and rags, etc. Many of these categories are often hazardous. However, in many cases the volumes of waste involved may be sufficiently small that it is cost effective to dispose of them as hazardous, rather than incurring the expense of testing for lead toxicity. This trade-off should be made by abatement contractors on a case-by-case basis."

Abatement contractors continued to experience uncertainties about disposal that were a deterrent to removal of lead-based paint. In 2001, the EPA adopted and then withdrew a complex rule on management and disposal of lead-based debris, as described in 66 Fed. Reg. 67108 (Dec. 21, 2001). In 2001, the EPA found a practical answer when it adopted amendments to its criteria for municipal solid waste landfills at 40 CFR Part 258. This was accomplished by amending the definition of a "municipal solid waste landfill unit" to exclude a construction and demolition landfill that receives residential lead-based paint waste and does not receive any other household waste. A construction and demolition (C&D) landfill may not receive hazardous waste, other than conditionally exempt small quantity generator waste, or industrial solid waste. The rule leaves open some questions but provides a destination perceived to be relatively economical for the bulk of lead-based paint demolition debris.

See Chapter XX, Subchapter I for further information about disposal of lead-based paint abatement waste.

X. COMPREHENSIVE ENVIRONMENTAL RESPONSE, COMPENSATION AND LIABILITY ACT (CERCLA)

A. General

The Comprehensive Environmental Response, Compensation and Liability Act (CERCLA) authorizes the EPA to take response actions to clean up facilities where there is a release or threatened release of a hazardous substance, pollutant or contaminant and to recover its response costs from the parties responsible for the contamination.[1] CERCLA also created a multibillion dollar trust fund, called the "Superfund," derived from excise taxes on oil and chemicals and other substances and intended for use by the EPA for response actions. The excise taxes are no longer in effect.

B. Release Reporting

CERCLA requires any person in charge of a vessel or facility to notify the National Response Center of any release of a reportable quantity of a hazardous substance from the vessel or facility.[2]

The EPA has listed lead and several lead compounds as hazardous substances and has assigned reportable quantities to most of them.[3]

Hazardous Substance	Reportable Quantity (RQ)
Lead	10 lb. (4.54 kg)
Lead acetate	10 lb. (4.54 kg)
Lead arsenate	1 lb. (0.454 kg)
Lead bis (acetato-0) tetrahydroxy tri	10 lb. (4.54 kg)
Lead chloride	10 lb. (4.54 kg)
Lead fluoroborate	10 lb. (4.54 kg)
Lead fluoride	10 lb. (4.54 kg)
Lead iodide	10 lb. (4.54 kg)

---

[1] 42 U.S.C. § 96D1 et seq.

[2] 42 U.S.C. § 9601.

[3] 40 CFR Part 302, Table 302.4.

Lead nitrate	10 lb. (4.54 kg)
Lead phosphate	10 lb. (4.54 kg)
Lead stearate	10 lb. (4.54 kg)
Lead subacetate	10 lb. (4.54 kg)
Lead sulfate	10 lb. (4.54 kg)
Lead sulfide	10 lb. (4.54 kg)
Lead thiocyanate	10 lb. (4.54 kg)

*No reporting is required if the diameter of solid metal pieces released equals or exceeds 100 micrometers (0.004 inches).

C. The National Contingency Plan; The Hazard Ranking System; and the National Priority List

The EPA does not necessarily undertake a response action merely because a release of hazardous substances is reported. The EPA has adopted a National Contingency Plan (NCP) containing a Hazard Ranking System (HRS) to determine the priorities. The HRS contains numerous criteria and a score of 28.5 is generally needed for listing of a facility on the National Priority List (NPL) for remedial action by the EPA. However, the EPA may take emergency removal actions without waiting to list a facility on the NPL.[4]

D. Performance of and Payment for Removal and Response Actions

The EPA does not usually perform removal actions or remediation of NPL facilities if responsible parties can be found who will consent, voluntarily or involuntarily, to perform the work. CERCLA authorizes the EPA to recover its response costs from present and former owners and operators of the facility and from persons who transported or arranged for transportation of hazardous substances to the facility. Since the liability is strict, retroactive and usually also joint and several, and available defenses are few and have been narrowly interpreted, the potentially responsible parties (PRPs) usually undertake the work. PRP may seek to recover

[4] 40 CFR Part 300 et seq.

contributions from other PRPs during or following any civil action under Sections 9606 or 9607(a) of CERCLA.[5]

E. No Hazard Threshold Levels Under CERCLA

It is important to recognize that there are no thresholds on the amount of a hazardous substance that can be the basis of CERCLA liability. If a substance is found to be hazardous, the fact that the quantity is less than the applicable RQ or less than the threshold required for a determination of hazardous waste under RCRA does not provide an excuse from liability.[6] Many persons who sent ordinary solid waste to municipal and other landfills have been required to pay large legal fees and liability settlements, although recently courts have granted relief to some defendants such as very small businesses.[7] Thus, abatement contractors and other persons who dispose of lead-containing waste must satisfy themselves that the disposal facility is unlikely to become the subject of a CERCLA cleanup.

F. Products in Use That Are Not Waste

CERCLA does not authorize response actions for a release or threatened release from products which are part of the structure of residential buildings or business or community structures. Accordingly, CERCLA does not authorize recovery of costs for removal of asbestos-containing materials from such buildings and structures.[8] The courts may reach the same result in lead-based paint cases.

---

[5] 42 U.S.C. §9613; Cooper Industries, Inc. v. Aviall Services, Inc., 543 U.S. 157 (2004).

[6] U.S. v. Alcan Aluminum Corp., 990 F.2d 711 (2d Cir. 1993); U.S. v. Alcan Aluminum Corp., 964 F.2d 252 (3d Cir. 1992).

[7] B.F. Goodrich v. Murtha, 815 F. Supp. 539 (D.C. Conn. 1993); later proceedings in the same case 840 F.Supp. 180 (D.C. Conn. 1994).

[8] First United Methodist Church v. U.S. Gypsum, 882 F.2d 862 (4th Cir. 1989) and cases cited therein.

G. Interim Guidance on Soil Cleanup

Once a facility is to be remediated under CERCLA, serious and controversial questions often arise about the extent of remediation required at a contaminated facility ("How Clean is Clean?").

In 1994, the EPA issued Office of Solid Waste and Emergency Response (OSWER) Directive #9355.4-12 to provide revised interim soil lead guidance for CERCLA sites and RCRA corrective action facilities. The Directive recommends a screening level of 400 ppm for lead in soil for residential land use. However, even if lead is below that level, further action might be required in special circumstances warranting further study. Above the 400 ppm level, a suggested decision logic is provided including evaluation of probable land use and exposure scenarios. The decision logic calls for collection of sampling data and blood level data for the specific site. The EPA recommends that case managers run the data in an IEUBK model to estimate risk and that key exposure pathways at the site be evaluated. If the IEUBK model shows that risks are significant, the EPA recommends that case managers evaluate remedial options. If lead-based paint is the only major contributor, no action is warranted because, as the EPA concedes, authority to remediate lead-based paint under CERCLA and RCRA is extremely limited. If soil is the only major contributor, a response to soil contamination is warranted, but paint abatement is not. If both exterior paint and soil are major contributors, remediation of both should be considered, particularly if deteriorated exterior paint will potentially recontaminate the soil. If indoor dust levels are greater than soil levels, the dust should be remediated by means such as HEPA equipment. Case managers are reminded that, although there is very little authority under CERCLA and RCRA to require cleanup of interior paint, they can consider whether the paint is being remediated for other reasons in their risk assessment for the site in a Record of Decision or Statement of Basis.

In late 1994, a citizens group filed a lawsuit challenging OSWER Directive #9355.4-12 on grounds that the EPA did not comply with the Administrative Procedures Act and that the standard and the IEUBK model lacked scientific basis.[9]

In 1998, the EPA issued OSWER Directives #9200.4-27P clarifying OSWER Directive #9355.412. OSWER pointed out that CERCLA requires it to select cleanup approaches that are protective of human health and the environment and that utilize permanent solutions to the maximum extent practicable. Blood lead studies can provide useful information to identify site-specific exposure pathways and to direct health professionals to individuals needing immediate assistance in minimizing lead exposure. However, blood lead studies do not attempt to identify risk factors for childhood exposure to lead sources and may not present an accurate representation of community-wide exposure. OSWER recommended that the IEUBK Model be used as the primary tool to generate risk-based soil cleanup levels at CERCLA and RCRA sites for current or future residential use. The levels should be designed to reduce risk to no greater than a 5% chance of exceeding a 10 ug/dL blood lead level for a full time child resident. OSWER said that response actions can be taken using IEUBK predictions alone and blood lead studies are not required for them. OSWER added that CERCLA provides limited authority to the EPA to address exposure from interior lead-based paint and limited resources to address exterior lead-based paint and soil contaminated by lead-based paint. Thus, OSWER recommended that EPA Regions promote remediation of those risks by others such as PRPs, local governments and homeowners.

OSWER and the EPA's Office of Prevention, Pesticides and Toxics (OPPT) sent a memorandum dated December 1, 1998 to the Regional Administrators stating that the 2,000 ppm hazard level for soils then proposed for rules under TSCA should not be treated as an Applicable or Relevant and Appropriate (ARAR), "to be considered" (TBC), or media cleanup standard (MCS). The IEUBK Model should be applied using site-specific information to

[9] Superfund Coalition v. U.S. Environmental Protection Agency, No. 94-1666 (D.C. Cir. Nov. 11, 1994).

evaluate risk at Brownfields, RCRA sites, NPL sites, state superfund sites, Federal CERCLA removal actions, and CERCLA non-NPL facilities.

H. Worker Protection under CERCLA

The EPA has provided worker protection to state and local government employees engaged in hazardous waste operations covered by CERCLA in states that do not have a state plan approved under the Occupational Safety and Health Act by making the substantive provisions of OSHA's HAZWOPER regulations at 29 §1910.120 applicable to them. The HAZWOPER regulations include requirements for protection against lead hazards.

I. Transfer of Federal Real Properties

The EPA has adopted a regulation under CERCLA requiring that any department, agency or instrumentality of the United States that enters into any contract for the sale or transfer of real property owned by the United States must include in the contract notice of the type, quantity and time of any hazardous substance stored for a year or more, known to have been released, or disposed of. 40 CFR Part 373.

XI. OCCUPATIONAL SAFETY AND HEALTH ACT

A. General

The Occupational Safety and Health Act (the OSH Act) created the Occupational Safety and Health Administration (OSHA) within the U.S. Department of Labor and established programs for the protection of safety and health in the nation's workplaces.[1] The OSH Act also established the National Institute for Occupational Safety and Health (NIOSH) to conduct research and experimental programs.

Major programs include adoption of safety and health standards; research programs including ways to discover latent diseases and work-related causes; hazard communication regulations requiring preparation of material safety data sheets (MSDS) and that employers assemble and make available to workers and their representatives information about hazardous materials used in the workplace; and nationwide inspections of workplaces to detect violations. The OSH Act requires each employer to furnish employment and a workplace free from recognized hazards causing or likely to cause death or serious physical harm. Both employers and employees must comply with the standards and the rules, regulations and orders of OSHA.

OSHA has adopted extensive regulations under the Act.[2] They include review and approval of state plans; inspections, citations and penalties; recording and reporting of occupational injuries; and safety and health standards. OSHA's General Industry Standard for Lead is at 29 CFR §1910.1025. The Construction Industry Standard is at 29 CFR§1926.62. The Shipyard Employment Standard is at 29 CFR §1915.1025.

B. The Hazard Communication Rule

OSHA has adopted hazard communication regulations requiring that employers develop, implement and maintain a written hazard communication program.[3] Each program must provide for communication to employees about hazardous chemicals and other hazards in the workplace by means such as labels and other forms of warning, material safety data sheets

---

[1] 29 U.S.C. § 651 et seq.

[2] 29 C.F.R. Part 1900 et seq.

[3] 29 CFR § 1910.1200.

(MSDSs) and employee information and training. A "chemical" is defined as any element, chemical compound or mixture of elements and/or compounds. A hazardous chemical is defined as any chemical which is a physical hazard or a health hazard. Lead is one of the chemicals listed as a health hazard.[4] Thus, chemical manufacturers and importers must label products containing lead and prepare and furnish MSDSs to their customers. The employers who receive the MSDSs must use them on their hazard communication programs.

An example of an MSDS for refined lead is provided later in this Chapter.

C. The General Industry Standard

1. Background

The General Industry Standard is a good example of the extensive, detailed systems which OSHA develops to regulate serious chronic health hazards when it determines that simpler systems are inadequate to achieve widespread effective worker protection.

Unlike imminent hazards, chronic hazards tend to be overlooked by some employers and employees, especially when the exposure is low level, intermittent and not readily obvious. This can lead to inadequate worker protection because some employers fail to provide adequate protection and some workers fail to use the protection provided. A simple regulatory program may be insufficient to overcome the inertia to an extent that effectively protects workers in the wide variety of workplaces that exist throughout the United States. When that happens, OSHA will adopt extensive detailed regulations which "spell out" the steps required to achieve an effective program. Employers must then comply with the detailed requirements or be subject to enforcement actions for violations. Employees must cooperate or be subject to discipline and perhaps even discharge by employers who cannot allow violations.

OSHA consulted with industry during its consideration of the Standard to determine its technological and economic feasibility. As a result of subsequent litigation, OSHA gave further consideration to several industries with special focus on the nonferrous foundry industry. OSHA granted a variance from the personnel exposure limit to that industry.

---

[4] 29 CFR § 1910.1000, Table Z-1.

The Standard imposes detailed requirements. As described later in this Chapter, OSHA and the courts, with a few exceptions, have interpreted the Standard in a broad and strict manner. They have ruled that compliance failures which some employers might believe are minor may be treated as serious and even egregious violations resulting in major penalties.

Although detailed, the Standard does not attempt to answer every question. Employers must often take voluntary initiatives in order to achieve compliance. Indeed, many employers succeed in achieving exposure reduction well beyond the requirements of the Standard. Some of the methods they use are discussed later in this Chapter.

OSHA has granted variances from the Standard. For example, OSHA granted variances to three automotive manufacturers for their solder grind booth operations, conditioned on their agreement to take several additional steps for protection of the affected employees including provision of positive pressure, supplied air respirators with hoods and protective bibs and more frequent monitoring and medical examinations. OSHA also granted a variance to a primary lead smelter from limits on the use of half-mask, air-purifying respirators equipped with high efficiency filters, but only for certain short-term tasks or under certain temperature extremes.

2. Summary of the General Industry Standard

This section summarizes key provisions of the General Industry Standard.

Scope and Application. The Standard applies to all occupational exposure to lead except for the construction industry and certain agricultural operations.[5] Lead is defined as metallic lead, all inorganic lead compounds and organic lead soaps, but not other organic lead compounds.

Permissible Exposure Limits (PELs) and Action Level.[6] Employers must assure that no employee is exposed to lead at concentrations greater than 50 micrograms per cubic meter of air (50 ug/m^3), based on an 8-hour time weighted average (TWA). This is the permissible exposure limit (PEL). The PEL is adjusted if an employee is exposed to lead for

[5] 29 CFR § 1910.1025.

[6] 29 CFR §1910.1025(c).

more than 8 hours on any workday. Protection provided by respirators may be considered in determining compliance for those periods the respirator is worn. The regulations also set a lower action level of 30 micrograms per cubic meter of air (30 ug/m^3) calculated as an 8-hour TWA.

Exposure Monitoring.[7] Every employer must make an initial determination by monitoring employee exposure to determine if any employee may be exposed to lead at or above the action level of 30 mg/m^3, 8-hour TWA without regard to respirator protection. The employer must base the initial determination on monitoring and any of the following relevant considerations (A) any information, observations or calculations indicating employee exposure to lead; (B) any previous measurements of airborne lead; and (C) any employee complaints of symptoms which may be attributable to lead exposure. Initial monitoring may be limited to a representative sample of employees exposed to the greatest concentrations of lead. NOTE: Many employers have never monitored for the presence of airborne lead because they operate businesses in buildings which they believe have no airborne lead. These employers are at risk of violation if they are mistaken.

If the initial determination is that no employee is exposed at or above the action level, the employer must make a written record of the determination and supporting information. The monitoring need not be repeated unless there is a production, process, control or personnel change or other reason to suspect a change which may result in new or additional exposure to lead.

If the initial determination reveals employee exposure at or above the action level but below the PEL, monitoring must be continued at least every six months until two consecutive measurements, taken at least 7 days apart, are below the action level. If the initial monitoring reveals that employee exposure is above the PEL, the employer must repeat monitoring quarterly until at least two consecutive measurements, taken at least 7 days apart, are below the PEL after which the employer must monitor at least every six months as described in the first sentence of this paragraph. Employers must notify each employee of his or her test results within 15

[7] 29 CFR §1910.1025(d).

working days after their receipt either individually in writing or by posting the results in an appropriate location accessible to employees.

Compliance Methods.[8] If an employee is exposed to lead above the PEL for more than 30 days per year, the employer must implement engineering and work practice controls, including administrative controls, to reduce and maintain employee exposure below the PEL except to the extent that the employer can demonstrate that such controls are not feasiable. Where they do not reduce employee exposure to or below the PEL, the employer must supplement the controls with respirators.

Each employer must establish and implement a written compliance program to reduce exposures to or below the PEL by means of engineering and work practice controls, including a detailed implementation schedule. If mechanical ventilation is used, the employer must evaluate its performance to maintain its effectiveness. If administrative controls are used, the employer must, among other steps, establish a job rotation schedule meeting specific requirements.

Respiratory Protection.[9] When required, employers must provide respirators that comply with several requirements. The selection requirements range from a half mask, air-purifying respirator with high efficiency filters for low exposures to full-face piece supplied-air respirators operated in positive pressure mode for high exposures. Employers must institute a respiratory protection program and must assure that respirators fit and are used properly. A checklist for respirator selection appears later in this Chapter. OSHA's standard for personal protective equipment, including respiratory protection, is at 29 CFR §1910.134.

Protective Work Clothing and Equipment.[10] If an employee is exposed to lead above the PEL without regard to the use of respirators, or where possible skin or eye irritation exists, the employer must provide at no cost (and assure that the employee uses) appropriate protective work clothing and equipment. The clothing may include coveralls or similar full body

---

[8] 29 CFR §1910.1025(e).

[9] 29 CFR §1910.1025(f).

[10] 29 CFR §1910.1025(f).

work clothing, gloves, hats, shoes or disposable coverlets. The equipment may include face shields, vented goggles or other items. The employer must provide for cleaning, laundering or disposal and for repair or replacement of protective clothing and equipment. The employer must also assure that protective clothing is removed only in change rooms and placed in closed containers to prevent dispersion of lead. Containers for contaminated clothing and equipment must have a warning label as follows: CAUTION: CLOTHING CONTAMINATED WITH LEAD. DO NOT REMOVE DUST BY BLOWING OR SHAKING. DISPOSE OF LEAD CONTAMINATED WASH WATER IN ACCORDANCE WITH APPLICABLE LOCAL, STATE, OR FEDERAL REGULATIONS.

Housekeeping.[11] Employers must maintain all surfaces as free as practicable of lead accumulations. Vacuum equipment with high efficiency particulate air (HEPA) filters are preferred. Compressed air is forbidden. Shovelling, dry or wet sweeping, and brushing may be used only where vacuuming or other equally effective methods have been tried and found ineffective. Vacuums must be used and emptied in a manner which minimizes reentry of lead into the workplace.

Hygiene Facilities and Practices.[12] Employers must assure that in areas where employees are exposed to lead above the PEL, regardless of respirator use, food or beverage is not present or used, tobacco products are not present or used, and cosmetics are not applied except in change rooms, lunch rooms or showers. Employers must provide clean change rooms equipped with separate storage facilities for protective clothing and equipment and for street clothes to prevent cross-contamination. Employers must provide shower facilities for employees whose airborne exposure is above the PEL without regard to respirators. Employers must also require that employees exposed to lead above the PEL (regardless of respirator use) take showers at the end of each shift and do not leave the workplace wearing any clothing or equipment worn during the work shift.

---

[11] 29 CFR §1910.1025(h).

[12] 29 CFR §1910.1025(i).

Employers must provide lunchroom facilities for employees who work in areas where their airborne exposure to lead is above the PEL without regard to the use of respirators. Employers must assure that these employees wash their hands and face before eating, drinking, smoking or applying cosmetics and do not enter lunchroom facilities or eating areas with protective clothing or equipment unless surface lead dust has been removed by vacuuming or other cleaning methods. Employers must also provide an adequate number of lavatories.

Medical Surveillance.[13] Employers must institute a medical surveillance program for all employees who are or may be exposed above the action level more than 30 days per year. All medical examinations and procedures must be performed by or under the supervision of a licensed physician. Employers must make available biological monitoring, consisting of blood sampling and analysis for lead and zinc protoporphyrin (ZPP) levels, to each employee covered by the program on a prescribed schedule. The sampling and analysis must meet frequency and accuracy specifications and be performed by a laboratory licensed by the Centers for Disease Control (CDC) of the Department of Health, Education and Welfare or which has received a satisfactory grade in blood lead proficiency testing from the CDC in the prior 12 months. Each employee whose blood lead level exceeds 40 ug/100 g of whole blood must be notified of his or her lead level and of the medical removal and benefits protection provisions of the Standard.

Employers must also make available medical examinations and consultations to each employee covered by the program. The required content of the medical examinations is extensive and includes a detailed work and medical history; a thorough physical examination; blood pressure measurement, blood analysis for blood lead level and other indicators; urinalysis; and any other test which the examining physician deems necessary by sound medical practice. Employers must notify employees of their right to have a second physician review findings or recommendations of the initial physician. If the two physicians do not agree and cannot quickly resolve their disagreement, then the employer and employee through their respective physicians shall designate and provide information to a third physician and the employer must act consistent with the findings, determinations and recommendations of the third physician unless the

---

[13] 29 CFR §1910.1025(j).

employer and employee reach an agreement consistent with the recommendations of at least one of the three physicians.

Employers must assure that any person they retain, employ, supervise or control does not engage in prophylactic chelation of any employee and that any therapeutic or diagnostic chelation is performed only under supervision of a licensed physician in a monitored clinical setting after prior written notice to the employee.

Medical Removal and Benefits Protection.[14] Employers must temporarily remove an employee from work having exposure to lead at or above the action level on each occasion when (A) a periodic and a follow-up blood sampling test indicates that the employee's blood level is at or above 60 ug/100g of whole blood, or (B) the average of the last 3 blood sampling tests or the average of all tests over the last 6 months, whichever is longer, indicates that the employee's blood lead level is at or above 50 ug/100 g of whole blood. However, an employee need not be removed if the last test is at or below 40 ug/100g of whole blood.

Employers must also temporarily remove an employee from such work whenever a final medical determination results in a medical finding, determination or opinion that the employee has a detected medical condition placing the employee at increased risk of material impairment of health from lead exposure.

The employer must return the employee to his or her former job status when two consecutive tests indicate that the employee's blood lead level is at or below 40 ug/100g of whole blood or, if the removal was based on a final medical determination, a subsequent final medical determination shows that the employee no longer has a detected medical condition of the kind resulting in the removal.

Employers must provide up to 18 months of medical removal protection benefits on each occasion when an employee is temporarily removed from exposure to lead or otherwise limited pursuant to the Standard. This means that the employer must maintain the employee's earnings, seniority and other employment rights and benefits as though the employee had not been removed. Employers may condition these benefits upon an employee's participation in

---

[14] 29 CFR §1910.1025(k).

medical surveillance made available pursuant to the Standard. Employers are entitled to reduce medical removal protection benefits to the extent that an award is made to an employee for earnings lost during the period of removal under workers compensation and other public or employer-funded programs. However, the employer is not entitled to any credit for workers compensation payments received by an employee for treatment related expenses.

The Standard requires employers to take specified measures if any employee's blood lead level does not decline within 18 months so that the employee is returned to his or her former job status. The measures include making available a medical examination to obtain a final medical determination whether the employee may be returned to his or her former job status and, if not, what steps should be taken to protect the employee's health. The employer must continue to provide medical removal protection benefits until either the employee is returned to former job status or a final medical determination is made that the employee is incapable of ever returning to his or her former job status. If an employer removes or limits an employee due to effects of lead exposure on the employee's medical condition, although not required by the Standard to do so, it must still provide medical protection benefits required by the Standard.

Employee Information and Training.[15] Each employer who has a workplace where there is potential exposure to airborne lead at any level must inform employees of the information in Appendices A and B to the Standard. If any employees are subject to exposure to lead at or above the action level or for whom the possibility of skin or eye irritation exists, the employer must provide for such employees a training program which must be repeated at least annually for each employee. The employer must also assure that each employee is informed of the content of the Standard and its Appendices and other more specific information. The Employer must also make readily available to all affected employees a copy of the Standard and its Appendices and must distribute to them any materials made available to the employer by OSHA.

[15] 29 CFR §1910.1025(l).

Signs.[16] Employers must post in each work area where the PEL is exceeded a warning sign as follows: "WARNING - LEAD WORK AREA - POISON - NO SMOKING OR EATING." They must assure that signs are illuminated and cleaned so that the legend is clearly visible.

Recordkeeping.[17] Employers must establish and maintain accurate records of all exposure monitoring, medical surveillance and medical removals and make them available, upon request, to employees, their designated representatives and OSHA. If an employer ceases to do business, any successor employer must receive and retain the records. If there is no successor employer, the employer must notify OSHA at least 3 months prior to disposal and deliver the records to OSHA if requested within that period.

Observation of Monitoring.[18] Employers must provide affected employees or their designated representatives an opportunity to observe monitoring of employee exposure to lead. Employers must furnish to the observers any necessary respirators, protective clothing and equipment and must require them to comply with applicable safety and health procedures. Observers are not entitled to interfere with the monitoring, but are entitled to an explanation of the measurement procedures and to observe all monitoring steps and record or receive copies of results.

3. Appendices to the General Industry Standard

The Standard includes several Appendices to assist in compliance. The Appendices are not intended to create any additional obligations or to detract from any existing obligation.

Appendix A - Substance Data Sheet. Appendix A contains basic information about lead and its compounds and their uses, the PEL and action level, and health hazard data written in a readable manner intended for distribution to employees.

---

[16] 29CFR §1910.1025(m).

[17] 29CFR §1910.1025(n).

[18] 29CFR §1910.1025(o).

Appendix B - Employee Standard Summary. Appendix B summarizes key provisions of the Standard in a readable manner intended for distribution to employees. It summarizes the provisions on the PEL, exposure monitoring, compliance methods, respiratory protection, protective work clothing and equipment, housekeeping, hygiene facilities and practices, medical surveillance, medical removal protection, employee information and training, signs, recordkeeping and observations of monitoring.

Appendix C - Medical Surveillance Guidelines. Appendix C outlines the medical surveillance provisions of the Standard. Part I explains the medical surveillance and monitoring requirements. Part II explains the adverse health effects of inorganic lead under the categories heme synthesis inhibition, neurological effects, gastrointestinal effects, renal effects, reproductive effects and other toxic effects. Part III describes methods used in medical evaluation. Part IV describes methods used in laboratory evaluation.

4. Enforcement of the General Industry Standard

Both OSHA and the courts have enforced the Standard strictly. In an early decision, the U.S. Court of Appeals for the District of Columbia upheld the Standard, although requiring partial reconsideration of several industries by OSHA.[19]

The courts and the OSHRC have generally rejected attempts to avoid or limit compliance. For example, an attempt to challenge the respirator requirements was rejected.[20] Benefits found to be less than the full benefits required by the temporary medical removal protection requirements were rejected by federal appellate courts, and the authority of OSHA to order backpay to abate such violations was upheld.[21] An employer which terminated benefits before the end of the 18-month temporary medical removal protection period because it determined that the workers could no longer work safely in a lead environment was found in

---

[19] United Steel Workers of America v. Marshall, 647 F.2d 1189 (D.C. Cir. 1980); cert. den. 453 U.S. 913 (1981). For later developments, see United Steelworkers v. Dole, 1987-1990 OSHD § 28,857 (D.C. Cir. 1990) and American Iron and Steel Institute v. OSHA, 939 F.2d 975 (D.C. Cir. 1991).

[20] RSR Corp. v. Donovan, 747 F.2d 294 (5 Cir. 1984).

[21] AFL-CIO, CLC v. St. Joe Resources Co., 916 F.2d 294 (5 Cir. 1990) and Dole v. East Penn Manufacturing Co., 894 F.2d 640 (3 Cir. 1990).

violation of the Standard.[22] Failure to comply with the frequency requirements for blood lead and ZPP tests was found to be a violation.[23] Failure to comply with medical examination frequency requirements was found to be a violation.[24] Failure to provide a lunchroom was found to be without an acceptable excuse.[25] Inadequate respirator programs have often been found to be serious violations.[26]

Employers have been excused from violations in some cases. An employer which unknowingly used lead-based paint in spray applications was excused.[27] An employer which used a sandblasting grit product for which the supplier furnished an MSDS that did not disclose lead was excused.[28] In an instructive case, the OSHRC closely reviewed a number of alleged violations cited by OSHA and affirmed by an Administrative Law Judge and upheld some, but vacated others.[29]

In a decision which received widespread public attention, a 5 to 4 majority of the U.S. Supreme Court ruled that a fetal protection policy of Johnson Controls, Inc., a battery manufacturer, excluding women of child-bearing capacity from work involving exposure over the action level facially violated Title VII of the Civil Rights Act of 1964 and that the manufacturer had not established a bona fide occupational qualification (BFOQ) defense. In view of the narrow majority, it may be significant to observe that three concurring Justices and the dissenting Justice believed that the majority had interpreted too narrowly the requirements to

---

[22] RSR Corp. v. Secretary of Labor, 764 F.2d 355 (5 Cir. 1985).

[23] Century Customs Corp., 1987-90 OSHD § 28,331.

[24] Advance Bronze, Inc., 1987-90 OSHD § 28,603.

[25] Central Brass Manufacturing Co., 1987-90 OSHD § 28,109.

[26] Advance Bronze, Inc. v. Dole, 917 F.2d 944 (6 Cir. 1990).

[27] Kut-Kwick Corp., 1983-84 OSHD § 26,796.

[28] Morrison-Knudson Industrial Services Co., 14 OSHC 1625, March 19, 1990.

[29] Atlantic Battery Co., Inc., 1995 OSHD § 30,636.

establish a BFOQ defense and should not have foreclosed such a defense based on consideration of potential tort liability for fetal injuries.[30]

During the 1990s, OSHA sought very large civil penalties for violations which it determined to be egregious. A striking example was approximately $1.5 million assessed against a company by the Occupational Safety and Health Review Commission.[31] (The Secretary of Labor had proposed penalties totalling $2,013,000.)

During 2001 and 2002, OSHA conducted a National Emphasis Program (NEP) for Lead during which it inspected over 700 establishments and issued over 1,800 citations for violations and announced that the NEP successfully reduced the severity of lead exposure in the workplace. OSHA said that it would continue the NEP and also local emphasis programs. For example, in 2003, OSHA proposed penalties totaling almost $50,000 against a foundry in Franklin, New Hampshire based on allegations of two willful, four repeated and 13 serious violations. OSHA proposed further penalties against the foundry in 2005.

Actions taken by OSHA to enforce its Construction Industry Standard are described later in this Chapter. In some cases, OSHA proposed multimillion dollar penalties for alleged violations that it characterized as egregious.

5. Working with the General Industry Standard
Management, Worker and Labor Union Support

Compliance with the Standard must be strongly, knowledgeably and consistently supported by management as a top priority. Although compliance is clearly an employer responsibility, worker and labor union cooperation is also vital.

Process and Engineering Controls

While the Standard speaks of engineering and work practice controls, OSHA recommends that employers also use process management methods to comply with the Standard and to exceed its requirements, if feasible. The effort should begin with selection and handling

---

[30] International Union, UAW v. Johnson Controls, Inc., 499 U.S. 197, 111 S. Ct. 1196 (1991).

[31] Interstate Lead Co. v. Secretary of Labor, 15 OSHC 1989 (Sept. 16, 1992).

of raw materials and the design or selection of facilities and equipment. It should continue throughout the process to assembly, packaging, loading and shipping of products. Whenever lead emissions, especially fugitive emissions, of lead can be reduced, there will be a reduced burden for remedial engineering control systems, administrative controls and personal protective equipment (PPE).

Among the remedial engineering controls, dust collection, air ventilation and exhaust, air filtration and recirculation, and vacuum surface cleaning are among the well known methods. Some companies used Hoffman central and "drop down" vacuum equipment. Some companies used power tools equipped with dust collection shrouds exhausted by vacuum through HEPA filtration equipment. Isolation of work areas where lead contamination is inevitable (for example, the use of clean booths) is an important engineering control method. Coating of floors and other surfaces to enhance the effectiveness of vacuuming is another useful method.

Work Practices and Personal Protective Equipment.

The Standard is very strict in its requirements for the use of PPE. This is a practical necessity because many employees dislike wearing respirators or other PPE. Although both comfort and effectiveness have been improved in recent years, respirators continue to be uncomfortable, especially when worn for long periods when the atmosphere is hot or humid, or both. However, OSHA has made it clear that management is responsible for proper selection, fit and use of respirators and will not ordinarily be excused from these responsibilities. Thus, management cannot excuse workers who do not use respirators when required.

The same principles apply to the use of change rooms, work clothes, showers, washing facilities and lunchrooms. Employers must provide the facilities, equipment, procedures and training. Employees must use them, even when it is inconvenient.

Company policies differ on the extent of PPE and other facilities which will be provided in excess of those required by the Standard. Some employers require employees to wear full face, powered air purifying respirators rather than half mask equipment, even when the Standard does not require it. Some require blood lead sampling and testing more frequently than required by the Standard. Some supply such equipment and personal vacuums, boot washers and smocks and foot covers for use in change rooms.

Health Programs

Health programs of experienced employers begin with the hiring process with health screening and education. Before they are hired, employees are given physical examinations to identify pre-existing health problems. The hazards of lead exposure and the requirements of the Standard and other safety and health programs are explained. Some companies require employees to acknowledge their training and agree in writing to cooperate with the health program.

Employers use various educational and motivational components in their programs. Some use periodic safety meetings including presentations by outside consultants with medical, industrial hygiene and related experience. Some hold an annual, offsite health and safety seminar to stimulate awareness. Some provide bonuses based wholly or partially on health-related goals such as reduced blood lead levels or maintenance of work areas in a clean condition so that air monitoring shows that respirators are not needed. For many employees, a "respirator free" work area is itself a welcome incentive.

For employees whose work necessarily involves exposure to lead, it is important to identify and minimize exposures that may occur both inside and outside the workplace. Thus, employers may monitor and restrict "hand to mouth" activities such as smoking and fingernail biting. Employers may ask workers to review their personal activities known to involve lead exposure such as paint removal, sports such as hunting or target practice, and making or use of pottery or ceramic ware which contain lead-based glazes.

A material safety data sheet (MSDS) for refined lead is provided later in this Chapter. A checklist of safety and health program elements is provided later in this Chapter.

When exposure is controlled, experience has shown that blood lead and ZPP levels generally decline as exposure is reduced. Thus, well planned health programs not only achieve compliance but also obtain measurably improved health effects.

6. Appendices for the General Industry Standard

A. Checklist for Respirator Selection

B. Material Safety Data Sheet for Refined Lead

C. Checklist of Health Program Elements

D. OSHA's Memorandum dated May 11, 2004 on Enforcement of the Engineering and Work Practice Control Provisions of the Lead Standard, 1910.1025(e)(1)

Appendix A

CHECKLIST FOR RESPIRATOR SELECTION

Airborne concentration of lead or condition of use	
Not in excess of 0.5 mg/m^3	Half-mask, air-purifying respirator equipped with high efficiency filters.[2,3]
Not in excess of 2.5 mg/m^3	Full facepiece air-purifying respirator with HEPA filters.[3]
	(1) Any powered air purifying respirator with HEPA filters; [3] or
	(2) Half-mask supplied-air respirator operated in positive-pressure mode. [2]
Not in excess of 100 mg/m^3	Supplied-air respirators with full facepiece, hood, helmet, or suit, operated in positive pressure mode.
Greater than 100 mg/m^3, unknown concentration or fire fighting.	Full facepiece, self-contained breathing apparatus operated in positive-pressure mode.

*Adapted from 29 CFR § 1910.1025, Table II

[1] Respirators specified for high concentrations can be used at lower concentrations of lead.

[2] Full facepiece is required if the lead aerosols cause eye or skin irritation at the use concentrations.

[3] A high efficiency particulate air (HEPA) filter means 99.97 percent efficiency against 0.3 micron size particles.

APPENDIX B

Minéraux
noranda
Minerals

Material Safety Data Sheet

WHMIS (Pictograms)	WHMIS (Classification)	Protective Equipment
	WHMIS CLASS D-2A: Material causing other toxic effects (VERY TOXIC).	

Section I: Chemical Product and Company Identification

Product Name/ Trade Name	Refined lead
Supplier	BRUNSWICK MINING AND SMELTING CORPORATION LIMITED Smelting and Fertilizer division Belledune, New Brunswick Canada E0B 1G0
Synonym	Lead metal, plumbum, plomb (French)
Chemical Name	Lead
Chemical Family	Metal. (Inert material/metal.)
Chemical Formula	Pb
Manufacturer	Refer to supplier.
Material Uses	Industrial applications: used principally in the battery industry.

Code	Not available.
CAS	7439-92-1
DSL	Not available.
CI	Not applicable.
In case of Emergency	(506) 522-2100 Extension: 211

Section II: Composition and Information on Ingredients

		Exposure Limits according to ACGIH			
Name	CAS #	TLV-TWA (mg/m³)	TLV-STEL (mg/m³)	TLV-CEIL (mg/m³)	% by Weight
Lead	7439-92-1	0.15			99.9

Toxicological Data on Hazardous Ingredients	Lead LD50: Not available. LC50: Not available.

Section III: Hazards Identification

Potential Acute Health Effects	Dangerous in case of ingestion and of inhalation. This product may irritate eyes and skin upon contact. As shipped this product does not present any special health hazards. Conditions and work practices which generate dust or fumes should be avoided or controlled. Dust and fumes may cause health effects.
Potential Chronic Health Effects	CARCINOGENIC EFFECTS: POSSIBLE HUMAN CARCINOGEN by EPA, IARC. MUTAGENIC EFFECTS: Not applicable. TERATOGENIC EFFECTS: SUSPECTED by OSHA. The substance may be toxic to blood, kidneys,nervous system, and reproductive system. A non-controlled repeated or prolonged exposure to this substance can produce target organ damage.

Section IV: First Aid Measures

Eye Contact	Check for and remove any contact lenses. Rinse eyes with running water for at least 15 minutes. Dust may act as a foreign body.
Skin Contact	Always wash with soap and water after handling refined lead and other lead products. Dust may irritate skin.

Continued on Next Page

Refined lead	***Page Number: 2***
Inhalation	**Remove patient from exposure and bring to fresh air. If breathing is difficult, give oxygen. Get immediate medical attention.** No immediate effects from dusts or fumes but inhalation is the main route of lead intoxication.
Ingestion	**Induce vomiting; refer to a physician. Do not induce vomiting or give a liquid to an unconscious person.** Immediate effects from fumes or dust are infrequent.

Section V: Fire and Explosion Data	
The Product is:	Non-flammable. Fine dust of lead will burn if ignited.
Auto-Ignition Temperature	Not applicable.
Flash Points	Not applicable.
Flammable Limits	Not applicable.
Products of Combustion	Lead oxide
Fire Hazards in Presence of Various Substances	Lead is flammable in the form of fine dust when exposed to heat or flames. Refer to special remarks on fire hazards.
Explosion Hazards in Presence of Various Substances	Slightly explosive in presence of open flames and sparks (in the form of dust). Risks of explosion of the product in presence of mechanical impact: dust explosion can occur in a confined space.
Fire Fighting Media and Instructions	Use dry chemical or CO2. Cool containers with water jet in order to prevent pressure build-up, autoignition or explosion.
Special Remarks on Fire Hazards	Lead is not flammable, however if in contact with a fire source lead will melt, and in contact with water will cause a violent reaction. When fire, possibility of toxic lead fumes. Firefighters should wear self-contained breathing apparatus (SCBA) and wear full protective clothing.
Special Remarks on Explosion Hazards	No additional remark.

Section VI: Accidental Release Measures	
Spill	Metal as shipped in pig or jumbo form can be recovered and accounted for, under normal clean up procedures for recycling.

Section VII: Handling and Storage	
Precautions	DO NOT ingest. DO NOT breathe dust. Wear suitable protective clothing. In case of insufficient ventilation, wear suitable respiratory equipment. If ingested, seek medical advice immediately and show the container, the label or the MSDS. Eating, drinking and smoking must be prohibited in area where this material is handled, stored and processed. Workers must wash hands and face before eating, drinking and smoking.
Storage	Store lead pigs and jumbos in covered, dry storage.

Continued on Next Page

APPENDIX B p.3

Refined lead	Page Number: 3

Section VIII Exposure Controls/Personal Protection

Engineering Controls	Use process enclosures, local exhaust ventilation, or other engineering controls to keep airborne levels below recommended exposure limits. If user operations generate dusts or fumes, use ventilation to keep exposure to airborne contaminants below the exposure limit.
Personal Protection	Safety glasses. Coveralls. Working gloves and boots. Dust respirator. Be sure to use a NIOSH/MSHA approved respirator or equivalent when occupational exposure limits are exceeded.
Personal Protection in Case of a Large Spill	If fumes are present or if dusts in large concentration are present, a self-contained breathing apparatus (SCBA) should be used to avoid inhalation of the product. If not in large concentration, use chemical goggles and a NIOSH/MSHA approved respirator or equivalent. Full suit. Boots. Gloves.
Exposure Limits	**Lead** TLV-TWA 0.15 (mg/m3) from ACGIH ACGIH notice of intended changes for 1993-1994 has proposed 0.05 mg/m3 for Lead elemental and inorganic compounds as Pb. These proposed changes to the TLV are only trial limits and they will remain on the list of TLVs for one year. Lead is also listed in it as A3: animal carcinogen. Consult local authorities for acceptable exposure limits.

Section IX Physical and Chemical Properties

Physical State and Appearance	Solid. (Soft and dense.)	Odor	Odorless.
Molecular Weight	207.19	Taste	Metallic
pH (1% soln/water)	Not available.	Color	Bluish-grey
Boiling Point	1744°C (3171.2°F)		
Melting Point	327.43°C (621.4°F)		
Critical Temperature	Not available.		
Specific Gravity	11.34 (Water = 1)		
Vapor Pressure	1mm @ 973°C		
Vapor Density	Not available.		
Volatility	Not applicable.		
Odor Threshold	Not applicable.		
Water/Oil Dist. Coeff.	Not applicable.		
Ionicity (in Water)	Not available.		
Dispersion Properties	Not available.		
Solubility	Very slightly soluble in cold water.		

Section X Stability and Reactivity Data

Stability	The product is stable.
Instability Temperature	Not applicable.
Conditions of Instability	Not applicable.
Incompatibility with various substances	See special remarks on reactivity.
Corrosivity	None

Continued on Next Page

APPENDIX B p.4

Refined lead	Page Number: 4
Special Remarks on Reactivity	**Molten lead metal** is a fire and explosion hazard and must be contained. When overheated, will emit toxic lead oxide fumes. Lead may react violently when used with certain compounds like: ammonium nitrate, chlorine trifluoride, hydrogen peroxide, sodium azide, sodium carbide, zirconium. If lead is used with these compounds, reference should be made to specific information on these reactions. Lead is incompatible with: sodium nitrate; zirconium; disodium acetylide; oxidants. Can react vigorously with oxidizing materials.
Special Remarks on Corrosivity	No additional remark.

Section XI: Toxicological Information	
Routes of Entry	Ingestion. Inhalation.
Toxicity to Animals	No additionel remarks. See section II.
Chronic Effects on Humans	**CARCINOGENIC EFFECTS:** POSSIBLE HUMAN CARCINOGEN by EPA, IARC. **MUTAGENIC EFFECTS:** Not applicable. **TERATOGENIC EFFECTS:** SUSPECTED by OSHA. The substance may be toxic to blood, kidneys,nervous system, and reproductive system. A non-controlled repeated or prolonged exposure to this substance can produce target organ damage.
Other Toxic Effects on Humans	Dangerous in case of ingestion and of inhalation. This product may irritate eyes and skin upon contact. As shipped this product does not present any special health hazards. Conditions and work practices which generate dust or fumes should be avoided or controlled. Dust and fumes may cause health effects.
Special Remarks on Toxicity to Animals	No additional remark.
Special Remarks on Chronic Effects on Humans	Lead is classified as a carcinogenic product: IARC and the EPA classified it as a group 2B carcinogen, possible carcinogen in humans. The ACGIH has classified it as a group A3 carcinogen, animal carcinogen. The Economic European Community Expert Committee on metals lists certain lead compounds as carcinogenic to animals. The committee does not list metallic lead as carcinogenic. Lead is listed as a teratogen and reproductive toxin. Lead is a regulated substance in many jurisdictions. NIOSH (90-117) reports the following target organs for acute and chronic overexposure: Gastro-intestinal tract, central nervous system, kidneys, blood, gingival tissues. Signs and symtoms of acute overexposure to lead usually develop slowly and resemble those of chronic overexposure which are: nauseas, vomiting, abdominal cramps, diarrhea, constipation, confusion, convulsions, anemia, muscular weakness and decreased sperm count. Acute overexposure is more likely to occur in children than in adults. Target organs for chronic overexposure are: blood, kidneys, digestive system, nervous system and the reproductive system..
Special Remarks on Other Toxic Effects on Humans	Workers with the following pre-existant conditions warrant particular attention: Anemia, pregnant women, breast feeding women and women of child bearing age. For biological monitoring, the preferred method is to measure blood lead levels (Pb blood). Eating, drinking and smoking must be prohibited in area where this material is handled, stored and processed. Workers must wash hands and face before eating, drinking and smoking.

Section XII: Ecological Information	
Ecotoxicity	Not applicable.
BOD5 and COD	Not applicable.
Products of Biodegradation	Not applicable.
Toxicity of the Products of Biodegradation	Not applicable.
Special Remarks on environment	No additional remark.

Continued on Next Page

APPENDIX B p.5

Refined lead	***Page Number: 5***

Section XIII: Disposal Considerations

Waste Disposal	Recycle to process, if possible. Consult the local or regional authorities.

Section XIV: Transport Information

TDG Classification	Not regulated under TDG (Canada).
PIN	Not applicable (PIN and PG).
Special Provisions for Transport	Not applicable.
TDG (Pictograms)	

Section XV: Other Regulatory Information and Pictograms

Other Regulations	OSHA: Hazardous by definition of Hazard Communication Standard (29 CFR 1910.1200).	
Other Classifications	HCS (U.S.A.)	HCS CLASS: DANGEROUS MAY CAUSE CANCER.
	DSCL (EEC)	39- Danger of very serious irreversible effects. 45- May cause cancer. 47- May cause birth defects.

Hazardous Material Information System (U.S.A.)

Health Hazard	2
Fire Hazard	0
Reactivity	0
Personal Protection	e

National Fire Protection Association (U.S.A.)

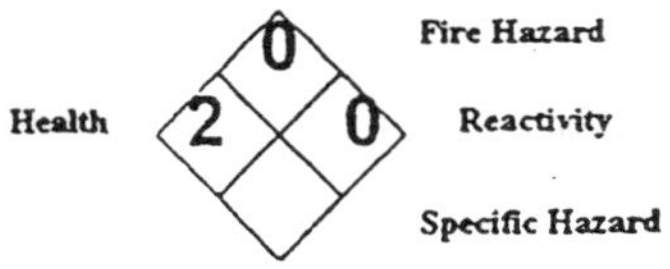

DOT (U.S.A) (Pictograms)

DSCL (Europe) (Pictograms)

ADR (Europe) (Pictograms)

Protective Equipment (Pictograms)

Continued on Next Page

APPENDIX B p.6

Refined lead		*Page Number: 6*
Section XVI: Other Information		
References	-NIOSH - RTECS, 1993 supplement. -ACGIH, Threshold Limit Values, Current listing. -Handbook of the Toxicology of Metals, 2nd Edition, Frieberg. -SAX'S Dangerous Properties of Industrial Materials, 8th Ed., 1993. -Patty's Industial Hygiene and Toxicology, 3rd Revised Edition. -National Library of Medicine -Toxicologie Ind. & Intox. Professionnelle, 3e édition, Lauwerys. -Chemical Hazards of the Workplace, Proctor, Hughes. -IARC, Monographs on the Evaluation of Carcinogenic Risks to Humans, Supp. 7. -Sixth Annual Report on Carcinogens, US, Public Health Service. -Toxicology Profile on selected metals. Contract: R. Lauwerys and Noranda Inc. (1989 with updates). -NIOSH - Pocket Guide to Chemical Hazards - June 1990. -CSST - Répertoire toxicologique, novembre 1991 (with updates). -ACGIH, documentation on TLVs and BEIs 1992-1993. -Controlled Products Regulation. IARC: International Agency for Research on Cancer. CSST: Commission de la Santé et de la Sécurité du Travail (Québec). ACGIH:American Conference of Governmental Industrial Hygiene. NTP: U.S. National Toxicology Program. NIOSH:National Institute of Occupational Safety and Health. OSHA:Occupational Safety and Health.	
Other Special Considerations	No additional remark.	
Validated on 93-11-22 by Paul Deveau.	**Verified by Dr. J-P. Robin.** **Printed 94-10-04.**	**Director, Occupational Health and Safety.**
Information Contact	**M.Paul Deveau, (506) 522-2100 ext. 367**	

Notice to Reader

Although reasonable precautions have been taken in the preparation of the data contained herein, it is offered solely for your information, consideration and investigation. Noranda Inc. extends no warranty and assumes no responsibility for the accuracy of the content and expressly disclaims all liability for reliance thereon. This material safety data sheet provides guidelines for the safe handling and processing of this product; it does not and cannot advise on all possible situations, therefore, your specific use of this product should be evaluated to determine if additional precautions are required. Individuals exposed to this product should read and understand this information and be provided pertinent training prior to working with this product.

Appendix C

CHECKLIST OF SAFETY AND HEALTH PROGRAM ELEMENTS

Statement of Policy

Notice of the Statement of Policy

Goals and Minimum Requirements

Role of Senior Management

Role of Safety, Health and Environmental Managers

Role of Engineering Managers in Engineering Controls

Role of Purchasing Managers and Suppliers

Role of Medical Personnel

Training of Managers and Supervisors Accountability

Delegation of Adequate Authority to Take Effective Action

Development of Standard Operating Procedures

Training of Employees and their Responsibilities

Program Implementation and Motivational Methods:

- Proper Facilities and Personal Protective Equipment
- Periodic Updated Training
- Safety and Health Meetings
- Bulletin Board or Newsletter
- Open Access for Employee Suggestions by Open Door Policy and/or Meetings
- Financial and Other Awards for Health-Related Achievements
- Individual Observation and Counselling

Inspections: Periodic and Surprise

Reporting of Violations and Problems

Discipline: Visibly Fair and Consistent

Recordkeeping

Standard Interpretations

05/11/2004 - Enforcement of the engineering and work practice control provisions of the lead standard, 1910.1025(e)(1).

Standard Interpretations - Table of Contents

- **Standard Number:** 1910.1025; 1910.1025(e)(1)

May 11, 2004

MEMORANDUM FOR: REGIONAL ADMINISTRATORS

FROM: RICHARD E. FAIRFAX, Director
DIRECTORATE OF ENFORCEMENT PROGRAMS

SUBJECT: Enforcement of the engineering and work practice controls provisions of the lead standard, 29 CFR 1910.1025(e)(1)

The Office of Health Enforcement (OHE) recently became aware of the need for clarification regarding the applicability of paragraph (e)(1) of the lead standard, 29 CFR 1910.1025[1]. This memorandum is to remind all field staff that the stay on enforcement of paragraph (e)(1) of the lead standard had been lifted for **all** industries. As you know, the courts had placed a stay of enforcement on paragraph (e)(1) for certain industries. These industries were non-ferrous foundries; secondary copper smelters; and independent collectors and processors of scrap lead, including battery breakers, leaded steelmaking operations, lead chemical manufacturers, and the brass and bronze ingot manufacturing. As of May 7, 2003, paragraph (e)(1) provisions are enforceable for all industries as follows:

Non-ferrous foundries: Those employers with >20 employees are required to achieve a lead level at the 50 $\mu g/m^3$ PEL. Smaller employers are required to achieve a lead level at 75 $\mu g/m^3$, which was determined to be the lowest feasible level for small foundries. However, these smaller employers must provide supplemental respiratory protection with an approved protection factor (APF) sufficient to reduce the 75 $\mu g/m^3$ level below the 50 $\mu g/m^3$ PEL for those employees in operations where the 50 $\mu g/m^3$ PEL cannot be achieved and maintained by engineering and work practice controls alone.

Secondary copper smelters: Employers are required to meet the PEL of 50 $\mu g/m^3$ through the use of engineering and work practice controls as is outlined in paragraph (e)(1).

Independent collectors and processors of scrap lead, including independent battery breakers: Employers are required to meet the PEL of 50 $\mu g/m^3$ through the use of engineering and work practice controls as is outlined in paragraph (e)(1).

Leaded steelmaking operations: Employers are required to meet the PEL of 50 $\mu g/m^3$ through the use of engineering and work practice controls as is outlined in paragraph (e)(1).

Lead chemical manufacturers: Employers are required to meet the PEL of 50 $\mu g/m^3$ through the use of engineering and work practice controls as is outlined in paragraph (e)(1).

Brass and Bronze Ingot Manufacturing: Employers are required to achieve and maintain an 8-hour TWA of 75 $\mu g/m^3$ solely by means of engineering and work practice controls. They must additionally provide supplemental respiratory protection (APF sufficient to be in compliance with the PEL) to each employee in every operation where the PEL of 50 $\mu g/m^3$ cannot be achieved and maintained by engineering and work practice controls alone.

Thank you for your attention to this matter. For further information refer to OSHA Instruction CPL 02-02-067 (formerly CPL 2-2.67). Please direct any inquiries to the Office of Health Enforcement at (202) 693-2190.

---

[1] Paragraph (e)(1) of the standard requires that, to the extent feasible, employers achieve the permissible exposure limit of 50 $\mu g/m^3$ solely by means of engineering and work practice controls. [Back to Text]

Standard Interpretations - Table of Contents

Back to Top www.osha.gov

Contact Us | Freedom of Information Act | Customer Survey
Privacy and Security Statement | Disclaimers

Occupational Safety & Health Administration
200 Constitution Avenue, NW
Washington, DC 20210

D. The Construction Industry Standard for Lead

This section summarizes key provisions of the Construction Industry Standard.

1. Background

Exposure to lead in the construction industry differs in several ways from the exposure in general industry. In general industry, workers in businesses such as lead smelting or battery manufacture will encounter potential exposure from the same sources for many years. Accordingly, development and consistent use of engineering controls, work practices and personal protective equipment can provide increasingly effective protection as years pass.

In the construction industry, even large projects have relatively brief duration. There is continuous change as work progresses from early steps, such as demolition and excavation, to final steps, such as painting and "punch list" items. A construction site may be used simultaneously by a number of owners, tenants, contractors, subcontractors, suppliers, transporters and other persons who interact with each other. Lead hazards may arise from unexpected sources and may affect employers and workers who do not have prior experience with lead. Employers and workers may be present at a site so briefly that blood lead test results are not known before their work is complete and they go to another construction project. For these and other reasons, OSHA allowed the construction industry, from 1979 to 1993, to comply with a PEL of 200 ug/m^3, 8-hour TWA that was four times as high as the PEL for general industry.

In the Housing and Community Development Act of 1992, Congress required that OSHA issue an interim final lead standard for the construction industry. The Act required that it be as protective as the worker protection guidelines for the identification and abatement of lead-based paint issued by the Department of Housing and Urban Development (HUD). In 1993, OSHA issued an interim final construction industry standard for lead without the customary notice, hearing and review procedures. The new standard contained provisions based on OSHA's general industry standard, HUD's guidelines for protection of workers engaged in abatement of lead-based paint, and recommendations of the Lead Work Group of the DOL's Construction Advisory Committee.

In adopting the Construction Industry Standard, OSHA consulted with members of the Lead Work Group of the DOL's Advisory Committee on Construction Safety and Health (ACCSH) which included representatives of management, labor and public health organizations and also with the Environmental Protection Agency. However, since the Act directed that OSHA adopt the Standard as an interim final rule within 180 days, OSHA could not take time for its customary rule making procedures such as publication of a proposed rule in the Federal Register followed by consideration of public comments presented in writing and perhaps at a public hearing.

The Construction Industry Standard parallels the General Industry Standard in structure and contains many identical requirements. However, it also contains numerous differences intended to reflect conditions in the construction industry and also to fulfill the requirement that it be at least as protective as HUD's worker protection guidelines. OSHA estimated that approximately 936,000 workers in 147,000 establishments were exposed to some level of lead during construction work and that the Construction Industry Standard would affect the entire construction industry as well as several nonconstruction industries.

2. Summary of the Construction Industry Standard[32]

Summaries of key provisions of the Construction Industry Standard are as follows:

Scope and Application. The Standard applies to all construction work where an employee may be occupationally exposed to lead. Lead is defined as metallic lead, all inorganic lead compounds and organic lead soaps, but not other organic lead compounds. Construction means all work for construction, alteration and/or repair, including painting and decorating. Examples of construction include such activities as demolition or salvage of structures containing lead or lead materials; installation of lead products; removal or encapsulation of lead-containing materials; and related onsite cleanup, transportation, disposal, storage, containment and maintenance activities.

[32] 29 CFR § 1926.62.

Permissible Exposure Limit (PEL) and Action Level.[33] Employers must assure that no employee is exposed to lead at concentrations greater than 50 micrograms per cubic meter of air (50 ug/m^3), based on an 8-hour time weighted average (TWA). This is the PEL. The PEL is adjusted if an employee is exposed to lead for more than 8 hours in any workday. Protection provided by respirators may be considered in determining compliance for those periods the respirator is worn. The regulations also set a lower action level of 30 ug/m^3 calculated as an 8 hour TWA.

Exposure Assessment.[34] Every employer must make an initial determination by monitoring employee exposure to determine if any employee may be exposed to lead at or above the action level of 30 ug/m^3, 8-hour TWA, without regard to respirator protection.

If the employer previously monitored for lead exposures and obtained data within the past 12 months during work operations conducted under workplace conditions closely resembling the processes, type of material, control methods, work practices, and environmental conditions used and prevailing in the employer's current operations, the employer may rely on the earlier monitoring data if the sampling and analytical methods meet required accuracy and confidence levels. NOTE: See the later discussion of enforcement proceedings for failure to perform initial monitoring in which employers unsuccessfully claimed reliance on prior monitoring data conducted under workplace conditions different from their current operations.

Employers must implement protective measures for employees performing certain lead related tasks which commonly produce lead exposures above the PEL until they perform the employee exposure assessment. These tasks include (A) manual demolition of structures, manual scraping, manual sanding, heat gun applications, and power tool cleaning with dust collection systems if lead containing coatings or paint are present and (B) and spray painting with lead paint. The tasks also include (A) using lead containing mortar and lead burning and (B) rivet busting, power tool cleaning without dust collection systems, cleanup activities where dry expandable abrasives are used, and abrasive blasting enclosure movement and removal if

---

[33] 29 CFR §1926.62(c).

[34] 29 CFR §1926.62(d).

lead containing coatings or paint are present. The tasks also include abrasive blasting, cutting and torch burning if lead containing coatings or paint are present on structures when they are performed. The tasks also include any others where the employer has any reason to believe that an employee may be exposed to lead in excess of the PEL unless the employer performs an employee exposure assessment and documents that the employee's lead exposure is not above the PEL.

Until the employer performs the required exposure assessment and determines actual employee exposure, the employer must provide to the employees performing the tasks respiratory protection; personal protective clothing and equipment; change areas; hand washing facilities; biological monitoring to consist of blood sampling and analysis for lead and ZPP levels; and training in subjects such as hazard communication, use of respirators, and safety training and education. NOTE: The interim protection requirements recognize the short-term nature of much construction work. If contractors could perform their work before performing an exposure assessment, many would complete most or all of the work before the exposure assessment could be completed.

If the initial determination reveals employee exposure to be below the action level, further determinations need not be repeated unless there is an equipment, process, control or process change or a new task that may result an employee being exposed above the action level.

If the initial determination or a subsequent determination is at or above the action level, monitoring must be continued until two consecutive measurements, taken at least 7 days apart, are below the action level. The monitoring frequency is six months if the employee exposure was at or above the action level but below the PEL. The frequency is 3 months if the employee exposure was above the PEL, but changes to 6 months when two consecutive measurements, taken at least 7 days apart, are at or below the PEL.

Employers must notify each employee of his or her test results as soon as possible but no later than 5 working days after their receipt either in writing or by posting the results in an appropriate location accessible to employees. Whenever the representative employee exposure was at or above the PEL without regard to respirators, the written notice must so state and describe the corrective action taken or to be taken to reduce exposure below the PEL.

Compliance Methods.[35] The employer must implement engineering and work practice controls, including administrative controls, to reduce and maintain employee exposure below the PEL to the extent such controls are feasible. Where they do not reduce employee exposure to or below the PEL, the employer must use the controls to reduce employee exposure to the lowest feasible level and supplement the controls with respirators.

Each employer must establish and implement a written compliance program to reduce exposures to or below the PEL, including a detailed implementation schedule. The program, which must be revised and updated at least annually, must include the following:

A. A description of each activity in which lead is emitted (e.g., equipment used, material involved, controls in place, crew size, employee job responsibilities, operating procedures, and maintenance practices);

B. A description of the specific means to achieve compliance where engineering controls are required; engineering plans and studies used to determine the methods selected for controlling exposure to lead;

C. A report of the technology considered in meeting the PEL;

D. Air monitoring data that document the source of lead emissions;

E. A detailed schedule for implementing the program, including documentation such as purchase orders for equipment, construction contracts, etc.;

F. A work practice program including use of protective work clothing and equipment and housekeeping and hygiene facility guidelines;

G. An administrative control schedule for job rotation, if used;

H. A description of arrangements made among contractors on multi-contractor sites to inform affected employees of potential exposure to lead and their responsibility to comply with this standard; and

---

[35] 29 CFR §1926.62(e).

If mechanical ventilation is used, the employer must evaluate its performance to maintain its effectiveness. The employer must ensure that, to the extent relevant, employees follow good work practices such as those described in Appendix B to the Construction Standard.

Respiratory Protection.[36] When required, employers must provide and assure the use of respirators. Respirators must be used during (1) periods when an employee's exposure to lead exceeds the PEL; (2) work operations for which engineering and work practice controls are insufficient to reduce employee exposures to or below the PEL; (3) periods when an employee requests a respirator; and (4) periods when respirators are required to provide interim protection while they perform the tasks for which interim protection is required until exposure assessment is performed.

Employers must select respirators based on the airborne concentration of lead and conditions of use. The requirements range from a half mask, air-purifying respirator with high efficiency filters for low exposures to full facepiece supplied-air respirators operated in positive pressure mode for high exposures. However, if an employee chooses to use a powered air-purifying respirator, the employer must provide such a respirator if it will provide adequate protection. Employers must implement a respiratory protection program in accordance with certain provisions of OSHA's standard for respiratory protection at 29 CFR §1910.134. A checklist for respirator selection appears later in this Chapter.

Protective Work Clothing and Equipment.[37] If an employee is exposed to lead above the PEL, without regard to the use of respirators, or if employees are exposed to lead compounds which may cause skin or eye irritation, the employer must provide at no cost to the employee, and assure that the employee uses, appropriate protective work clothing and equipment. The clothing may include (1) coveralls or similar full body work clothing; (2) gloves, hats, shoes or disposable shoe coverlets; and (3) face shields, vented goggles or other items. The employer must provide cleaning, replacement, change rooms and closed containers to prevent dispersion of lead. Containers for contaminated clothing and equipment must have

---

[36] 29 CFR §1926.62(f).

[37] 29 CFR §1926.62(g).

warning labels. The employer must inform cleaners and launderers about the potentially harmful effects of lead exposure.

Housekeeping.[38] Employers must maintain all surfaces as free as practicable of lead accumulations. If vacuuming methods are selected, vacuum equipment with high efficiency particulate air (HEPA) filters is required. Compressed air is forbidden unless used with a ventilation system designed to capture the airborne dust created. Shovelling, dry or wet sweeping, and brushing may be used only where vacuuming or other equally effective methods have been tried and found ineffective.

Hygiene Facilities and Practices.[39] Employers must assure that in areas where employees are exposed to lead above the PEL, regardless of respirator use, there are no food, beverages or tobacco products and cosmetics are not applied. Employers must provide clean change areas equipped with separate storage facilities for protective clothing and equipment and for street clothes to prevent cross-contamination. Employers must provide shower facilities, where feasible, for employees whose airborne exposure is above the PEL, regardless of respirator use. Employers must require that employees take showers at the end of each workshift and do not leave the workplace wearing any clothing or equipment worn during the work shift. Employers must provide lunchroom facilities or eating areas as free as practicable from lead contamination and readily accessible to employees. Employers must assure that employees exposed to lead above the PEL (regardless of respirator use) wash their hands and face before eating, drinking, smoking or applying cosmetics and do not enter lunchroom facilities or eating areas with protective work clothing or equipment unless surface lead dust has been removed by vacuuming or other cleaning method. Employers must provide hand washing facilities and, where showers are not provided, assure that employees wash their hands and face at the end of the workshift.

---

[38] 29 CFR§1926.62(h).

[39] 29 CFR§1926.62(i).

Medical Surveillance.[40] Employers must make available initial medical surveillance, consisting of blood sampling and analysis for lead and ZPP levels, to employees occupationally exposed on any day to lead at or above the action level, at no cost to such employees. Employers must also institute a medical surveillance program for all employees who are or may be exposed above the action level more than 30 days in any consecutive 12 months. All medical examinations and procedures must be performed by or under the supervision of a licensed physician. Employers must make available biological monitoring, consisting of blood sampling and analysis for lead and zinc protoporphyrin (ZPP) levels, to each employee covered by the program, and the sampling and analysis must be repeated on prescribed schedules until two consecutive samples indicate that his or her blood lead level is below the action level.

The frequency for employees exposed above the action level, but below the PEL, is every 2 months for the first 6 months and every 6 months thereafter. The frequency for employees exposed at or above the PEL is every 2 months until two consecutive blood samples and analyses indicate a blood lead level below 40 ug/dL.

The sampling and analysis must meet accuracy specifications and be performed by a laboratory approved by OSHA. Each employee whose blood lead level exceeds 40 ug/dl must be notified of his or her lead level and about the medical removal and benefits protection provisions of the Standard.

Employers must also make available medical examinations and consultations to each employee covered by the program. The required content of the medical examinations is extensive and includes a related work history and medical history; a thorough physical examination; blood pressure measurement; blood analysis for blood lead, ZPP and other components; urinalysis; several other specific tests; and any other test which the examining physician deems necessary by sound medical practice. Employees are entitled to initiate a multiple physician review.

Employers must assure that any person for whom they are responsible does not engage in prophylactic chelation, and performs therapeutic or diagnostic chelation only under

---

[40] 29 CFR§1926.62(j).

supervision of a licensed physician in a monitored clinical setting after prior written notice to the employee.

Medical Removal and Benefits Protection.[41] Employers must temporarily remove an employee from work having exposure to lead at or above the action level on each occasion when a periodic and a follow-up blood sampling test indicates that the employee's blood level is at or above 50 ug/dL. Employers must also temporarily remove an employee from such work whenever a final medical determination results in a medical finding, determination or opinion that the employee has a detected medical condition placing him or her at increased risk of material impairment of health from lead exposure. The employer must return the employee to his or her former job status when additional sampling and testing shows that his or her blood levels have declined to a level at or below 40 ug/dL, if blood lead level was the reason for removal, or a subsequent final medical determination shows that the employee no longer has a detected medical condition of the kind resulting in his or her removal.

Employers must provide up to 18 months of medical removal protection benefits on each occasion when an employee is temporarily removed from exposure to lead or otherwise limited pursuant to the Standard. This means that the employer must maintain the employee's earnings, seniority and other employment rights and benefits as though the employee had not been removed. Employers may condition these benefits upon an employee's participation in medical surveillance made available pursuant to the Standard. Employers are entitled to reduce medical removal protection benefits to the extent that an award is made to an employee for earnings lost during the period of removal under workers compensation and other public or employer-funded programs. However, an employer is not entitled to any credit for workers compensation payments received by an employee for treatment related purposes.

The Standard requires employers to take specified measures if any employee's blood lead level does not decline within 18 months so that the employee has been returned to his or her former job status.

---

[41] 29 CFR§1926.62(k).

Employee Information and Training.[42] The employer must communicate information about lead hazards to all employees who may be exposed to lead (regardless of actual exposure level) according to OSHA's Hazard Communication Standard for the construction industry (29 CFR § 1926.59) including warning signs and labels, material safety data sheets (MSDSs), information and training. If any employees are subject to exposure at or above the action level on any day or to lead compounds which may cause skin or eye irritation, the employer must provide and assure participation of all such employees in a training program which must be repeated at least annually for each employee who is subject to lead exposure at or above the action level on any day. The training program must include eight specific items listed in the Standard. The employer must also assure that each employee is informed of the content of the Standard and its Appendices and other more specific information. The Employer must also make readily available to all affected employees a copy of the Standard and its Appendices and must distribute to them any materials made available to the employer by OSHA.

Signs.[43] Employers must post in each work area where the PEL is exceeded a warning sign as follows: "WARNING - LEAD WORK AREA - POISON - NO SMOKING OR EATING." They must assure that signs are illuminated and cleaned as necessary so that the legend is clearly visible. Employers must also assure that no statement appears on or near any required sign that contradicts or detracts from its meaning.

Recordkeeping.[44] Employers must establish and maintain accurate records of exposure assessments, medical surveillance, medical removals and objective data for any exemption from the initial monitoring requirement and make them available, upon request, to employees, former employees and their designated representatives and to OSHA. If an employer ceases to do business, any successor employer must receive and retain the records. If there is no successor employer, the employer must notify OSHA at least 3 months prior to disposal and deliver the records to OSHA if requested within that period.

---

[42] 29 CFR§1926.62(l).

[43] 29 CFR§1926.62(m).

[44] 29 CFR§1926.62(n).

Observation of Monitoring.[45] Employers must provide affected employees or their designated representatives an opportunity to observe monitoring of employee exposure to lead conducted pursuant to the monitoring requirements of the standard. Employers must furnish to the observers any necessary respirators, protective clothing and equipment and must require them to comply with applicable safety and health procedures. Observers are not entitled to interfere with the monitoring, but are entitled to an explanation of the measurement procedures and to observe all monitoring steps and record or receive copies of results.

3. Appendices to the Construction Industry Standard

The Standard includes several Appendices to assist in compliance. The Appendices are not intended to create any additional obligations or to detract from any existing obligation.

Appendix A - Substance Data Sheet. Appendix A contains basic information about lead and its compounds, their uses in construction, the PEL and action level, and health hazard data written in a readable manner intended for distribution to employees.

Appendix B - Employee Standard Summary. Appendix B summarizes key provisions of the Standard in a readable manner intended for distribution to employees. It summarizes the provisions on the PEL, exposure monitoring, compliance methods, respiratory protection, protective work clothing and equipment, housekeeping, hygiene facilities and practices, medical surveillance, medical removal protection, employee information and training, signs, and observations of monitoring.

Appendix C - Medical Surveillance Guidelines. Appendix C outlines the medical surveillance provisions of the Standard. Part I explains the medical surveillance and monitoring requirements. Part II explains the adverse health effects of inorganic lead under the categories heme synthesis inhibition, neurological effects, gastrointestinal effects, renal effects, reproductive effects and other toxic effects. Part III describes methods used in medical evaluation. Part IV describes methods used in laboratory evaluation.

---

[45] 29CFR§1926.62(o).

4. Enforcement of the Construction Industry Standard

OSHA began its enforcement of the construction industry with several widely publicized enforcement actions seeking precedent-breaking civil penalties for violations alleged to be egregious. In February 1, 1994, OSHA proposed civil penalties totalling $1,300,000 against Manganas Paint Co. for alleged egregious violations in sandblasting a bridge in Ohio. On June 3, 1994, OSHA proposed penalties totalling $5,000,000 against E. Smalis Painting Co. for alleged egregious violations. On December 14, 1994, OSHA proposed additional penalties totalling $2,400,000 against Manganas Paint Co. for alleged egregious violations.

In proceedings during the 1990s in which egregious violations were not alleged, OSHA proposed large civil penalties totalling $154,300 against Dynamic Painting Corp., a general contractor, and $46,200 against Team Star Contractors, Inc., a subcontractor, for alleged violations while repainting a bridge in Massachusetts. OSHA also proposed civil penalties totalling $293,500 against M.P. Industries, Inc. for 18 violations while working on the Old Executive Office Building which is part of the White House Complex in Washington, D.C. Most of the violations were classified as willful or serious, but not as egregious. OSHA also proposed civil penalties totalling $97,500 against M.L. Young Painters for 16 serious violations.

In 2001, OSHA proposed civil penalties totaling $250,000 for alleged willful violations and serious violations by Lunda Construction Co. while demolishing a bridge in Milwaukee, Wisconsin.

In 2005 and 2006, OSHA charged several contractor firms with failure to conduct initial exposure assessments that would have revealed the presence of lead-based paint and potential hazards. OSHA also charged the firms with violations that resulted from the failure to perform an initial assessment such as failure to provide protective clothing and a clothing change area, failure to provide training, and failure to monitor blood lead levels. Thus, the potential penalties for willful and serious violations became quite significant. Some contractors said that they did not conduct an exposure assessment based on a similar previous project, but failed to show a close resemblance of workplace conditions. Bianchi v. Chao, 21 OSHC 1113, 3d Cir., June 1, 2005 (Demolition of Three Rivers Stadium); Techno Coating v. OSHRC, 21 OSHC

1473, 9th Cir., April 19, 2006 (Paleta Creek Pier Naval Base); Shaw Steeple Jacks, March 7, 2006 (Proposing $114,750 of civil penalties); and Leo J. Roth Corp., April 10, 2006 (Proposing $323,000 of penalties).

For additional background, see the discussion of OSHA's enforcement of the General Industry Standard earlier in this Chapter.

5. Working with the Construction Industry Standard

(a) General Construction Activities

OSHA has issued a variety of publications which are helpful in achieving compliance including "Lead in Construction," OSHA 3142, 1993 and explanatory Fact Sheets.

Many of the methods used to comply with OSHA's General Industry Standard can also be used to comply with the Construction Industry Standard. See also the discussion earlier in this Chapter for practical recommendations on working with the General Industry Standard.

By letters dated March 1, 1999 and May 8, 2000, OSHA indicated that it did not consider XRF analysis to be an acceptable method of analysis to determine employee airborne exposure. OSHA acknowledged that XRF analyzers are an acceptable method for meeting HUD/EPA requirements, but said that its concerns are different from those of HUD and EPA. OSHA does not consider any method that relies solely on analysis of bulk materials or surface content to be acceptable without air monitoring results which determine the airborne concentration of lead.

(b) Abatement of Lead-Based Paint by the Construction Industry

Although the Construction Industry Standard has broad application to construction activities, the primary concern which led Congress to require OSHA to adopt the new standard was the growing volume of work involving the identification and abatement of lead-based paint.

Lead-based paint is often identified by swab tests or by a portable direct reading x-ray fluorescence (XRF) analyzer which can be used with reasonable reliability for preliminary screening. Where more precise analysis is needed, paint chip samples can be taken and sent for analysis at a qualified laboratory using flame atomic absorption spectrophotometric (FAAS) equipment or graphite furnace atomic absorption spectrophotometric (GFAAS) equipment.

Lead-based paint is abated by removal, encapsulation or management in place. Removal can be accomplished by removing the painted component such as a door or window frame. It can also be done by removing the paint from surfaces by such means as chipping, stripping or abrasive blasting, onsite or offsite. Encapsulation can be done by proper application of a permanent protective surface such as sheetrock or plywood.

Numerous federal, state and municipal laws apply to abatement and require, prohibit or restrict abatement methods. For example, methods such as dry chipping and sanding, abrasive blasting, torch or open flame, machine sanding, and use of methylene chloride are prohibited or restricted. Encapsulating by epoxy or other durable coatings may be unacceptable to environmental agencies who believe these coatings might not provide sufficient protection to justify their use, although they offer the advantages of low cost and ease of application.

In abatement of lead-based paint, OSHA recommends a variety of engineering controls. They include (1) exhaust ventilation, (2) enclosure and encapsulation, (3) substitution of materials and equipment, (4) replacement of building components rather than their abatement by methods which create contamination, (5) modification of processes and equipment including the addition of HEPA filtration equipment, and (6) isolation of operations which create contamination. A copy of OSHA's Fact Sheet No. 93-48 discussing engineering controls appears as Appendix B at the end of this Chapter.

The Society for Protective Coatings (SSPC), 40 24th Street, Pittsburgh, PA 15222, (412) 281-2331; the National Association of Corrosion Engineers (NACE), 1440 S. Creek Drive, Houston, Texas 77084-4906, (800) 797-6223; and the International Safety Equipment Association (ISEA), 1901 N. Moore Street, Arlington, Virginia 22209 (703) 525-1695, are among the industry organizations which have publications or other information helpful for abatement of lead-based paint.

(c) Welding, Cutting and Heating Regulations

OSHA's regulations for the construction industry include requirements for ventilation and protection for welding, cutting and heating in confined spaces or in enclosed spaces involving metals of toxic significance including lead-based metals and metals containing lead, other than as an impurity, or metals coated with lead-bearing materials. Among other things, mechanical ventilation must have sufficient capacity and be so arranged as to keep fumes and smoke within safe limits, properly discharge contaminated air, and provide clean and responsible replacement air.[46]

When welding, cutting or heating any surface covered with a toxic preservative coating in an enclosed space, the surface covered with that coating must be stripped of all toxic coatings for a distance of at least 4 inches from the area of heat application or the employees must be protected by air line respirators meeting OSHA requirements. In open air, employees must be protected by a respirator meeting OSHA's requirements.[47]

(d) Review of the Construction Industry Standard

On June 6, 2005, OSHA published notice of a review of the Construction Industry Standard purportedly to determine whether there are ways to make implementation more practical, to reduce regulatory burden on small business, and improve its effectiveness while still protecting worker health.[48] The notice dramatically described the adverse health effects of lead, including incidents in 1988, 1992 and 1994 when bridge workers were found to have blood lead levels considerably exceeding the PEL. The notice also said that lead-based paint continued to be widely prevalent in housing and on structural steel bridges, storage tanks, sports stadiums and pipelines. OSHA requested comments, including answers to a list of questions by September 2005 and later extended the comment period. OSHA expects to announce the results of the review during 2006.

---

[46] 29 CFR §1926.353.

[47] 29 CFR §1926.354.

[48] Fed. Reg. Vol. 70, p. 32739, June 6, 2005.

6. Appendices for the Construction Industry Standard

E. Checklist for Respirator Selection

F. OSHA's Fact Sheet No. 93-48 on Engineering Controls.

G. OSHA's Memorandum dated 12/08/98 on Enforcement Policy for Abrasive-Blasting Respiratory Protection under the Lead in Construction Interim Final Rule

NOTE: Subchapter C - The General Industry Standard and Subchapter D - The Construction Industry Standard in this Chapter were obtained from chapters written by the author for Workplace Safety and Health, Thompson Professional Publishing (1995) and were included with its consent. They were updated for the Second Edition.

<u>Appendix E</u>

CHECKLIST FOR RESPIRATOR SELECTION*

Airborne concentration of lead or condition of use	Required respirator[1]
Not in excess of 500 ug/m^3	• ½ mask air purifying respirator with high efficiency filters.[2,3] • ½ mask supplied air respirator operated in demand (negative pressure) mode.
Not in excess of 1,250 ug/m^3	• Loose fitting hood or helmet powered air purifying respirator with high efficiency filters.[3] • Hood or helmet supplied air respirator operated in a continuous-flow mode, e.g.,
Not in excess of 2,500 ug/m^3	• Full facepiece air purifying respirator with high efficiency filters.[3] • Tight fitting powered air purifying respirator with high efficiency filters.[3] • Full facepiece supplied air respirator operated in demand mode. • ½ mask or full facepiece supplied air respirator operated in continuous-flow mode. • Full facepiece self-contained breathing apparatus (SCBA) operated in demand mode.

Airborne concentration of lead or condition of use		Required respirator[1]
Not in excess of 50,000 ug/m^3	•	½ mask supplied air respirator operated in pressure demand or other positive-pressure mode.
Not in excess of 100,000 ug/m^3	•	Full facepiece supplied air respirator operated in pressure demand or other positive-pressure mode, e.g., type CE abrasive blasting respirators operated in a positive-pressure mode.
Greater than 100,000 ug/m^3 unknown concentration, or fire fighting	•	Full facepiece SCBA operated in pressure demand or other positive-pressure mode.

*Adapted from 29 CFR § 1926.62, Table I

[1]Respirators specified for higher concentrations can be used at lower concentrations of lead.

[2]Full facepiece is required if the lead aerosols cause eye or skin irritation at the use concentrations.

[3]A high efficiency particulate filter air (HEPA) means a filter that is 99.97% efficient against particles of 0.3 micron size or larger.

OCCUPATIONAL SAFETY AND HEALTH ACT

U.S. Department of Labor Program Highlights

Fact Sheet No. OSHA 93-48
Lead Exposure in Construction
(#2 in a series of 6)

Engineering Controls

Because lead is a cumulative and persistent toxic substance and because lead-caused health effects may result from low levels of exposure over prolonged periods of time, engineering controls and good work practices must be used where feasible to minimize employee exposure to lead. At a minimum, exposures must not exceed the OSHA interim final PEL of 50 micrograms per cubic meter of air (50 $\mu g/m^3$) averaged over an 8-hour period. When feasible engineering controls and work practice controls cannot reduce worker exposure to lead to at or below 50 $\mu g/m^3$, respirators must be used to supplement the use of engineering and work practice controls.

A competent person should review all site operations and stipulate the specific engineering controls and work practices designed to reduce worker exposure to lead. Engineering measures include local and general exhaust ventilation, process and equipment modification, material substitution, component replacement, and isolation or automation. Examples of recommended engineering controls that can be used to reduce worker exposure to lead are as follows:

- **Exhaust Ventilation**

Power tools used for the removal of lead-based paint should be equipped with dust collection shrouds or other attachments exhausted through a high-efficiency particulate air (HEPA) vacuum system. Operations such as welding, cutting/burning, heating should be provided with local exhaust ventilation. HEPA vacuums should be used during clean-up activities.

For abrasive blasting operations where full containment exists or is required, the containment structure should be designed to optimize the flow of ventilation air past the worker(s), so that the airborne concentration of lead is reduced and the visibility is increased. The affected area should be maintained under negative pressure to reduce the chances that lead dust will contaminate areas outside the enclosure. A containment structure should be equipped with dust collection and an air-cleaning device to control emissions of particulate matter to the environment.

- **Enclosure/Encapsulation**

Lead-based paint can be made inaccessible either by encapsulating it with a material that bonds to the surface, such as acrylic or epoxy coating or flexible wall coverings, or by enclosing it using systems such as gypsum wallboard, plywood panelling, and aluminum, vinyl or wood exterior siding. Floors coated with lead-based paint can be covered using vinyl tile or linoleum flooring.

The building owner, or other responsible person, should oversee the custodial and maintenance staffs and contractors with regard to all activities that involve enclosed or encapsulated lead-based paint. This will minimize potential inadvertent release of lead during maintenance, renovation, or demolition.

- **Substitution**

Zinc-containing primers covered by an epoxy intermediate coat and polyurethane topcoat are commonly used instead of lead-containing coatings.

Mobile hydraulic shears can be substituted for torch cutting under certain circumstances.

Surface preparation equipment, such as needle guns with multiple reciprocating needles completely enclosed within an adjustable shroud, can be substituted for abrasive blasting under certain operations. The shroud captures dust and debris at the cutting edge and can be equipped with a HEPA vacuum filtration system with a self-drumming feature. One such commercial unit can remove lead-based paint from flat steel and concrete surfaces, outside edges, inside corners, and pipes.

Chemical strippers used primarily on the exterior of buildings, surfaces involving carvings or molding, or intricate iron works, can be used in place of hand scraping using a heat gun. Chemical removal generates less airborne lead dust.

These strippers, however, can be hazardous and the material safety data sheets (MSDSs) for the products used must be reviewed by the employer for information on worker exposure hazards from the chemical ingredients and protective measures recommended by the manufacturer.

- **Component Replacement**

Lead-based painted building components (i.e., windows, doors, and trim) can be replaced either with new components free of lead-containing paint or with the same components after the paint has been removed off-site. Replacement is a permanent solution.

- **Process/Equipment Modification**

Brush/roller application of lead paints or other lead-containing coatings is a safer method than spraying. (Note: There is a ban on the use of lead-based paint in residential housing.) This method of application introduces little or no paint mist into the air where the mist can present a lead inhalation hazard.

Non-silica containing abrasive (e.g., steel or iron shot/grit) should be used where practical instead of sand in abrasive blasting operations. The free silica portion of the dust presents a respiratory health hazard.

Blasting techniques that are less dusty than abrasive blasting and that can be effective under some conditions include: (1) hydro- or wet-blasting (using high pressure water with or without abrasive or surrounding the blast nozzle with a ring of water), and (2) vacuum blasting where a vacuum hood for material removal is positioned around the exterior of the blasting nozzle.

Heat guns used to remove lead-based paints in residential housing units should be of the flameless electrical softener type. Heat guns should have electronically controlled temperature settings to allow usage below 700°F. Heat guns should be equipped with various nozzles to cover all common applications and to limit heated work area.

When using abrasive blasting with vacuum on exterior building surfaces, care should be taken that the configuration of the heads on the blasting nozzle match the configuration of the substrate so that the vacuum is effective in containing debris.

Since HEPA vacuum cleaners can be used to clean surfaces other than just floors, operators should have attachments appropriate for use on unusual surfaces. The proper use of brushes of various sizes, crevice tools and angular tools, when needed, will enhance the quality of the HEPA-vacuuming process and help reduce the amount of lead dust released into the air.

- **Isolation**

Although it is not feasible to completely enclose and ventilate some abrasive blasting operations, it is possible to isolate many operations to help reduce the potential for exposure to lead. Isolation, in this instance, consists of keeping employees not involved in the blasting operations as far away as possible from the work area. By placing the employees at a greater distance from the source of lead exposure, their exposures will be reduced.

# # #

Standard Interpretations

12/08/1998 - Enforcement policy for abrasive-blasting respiratory protection under the Lead in Construction Interim Final Rule.

Standard Interpretations - Table of Contents

- **Standard Number:** 1926.62(f); 1910.134

This memorandum provides specific enforcement policy for respiratory protection required in abrasive-blasting operations under the Interim Final Rule for Lead in Construction, 29 CFR 1926.62 (hereafter called the "Lead in Construction Standard"). Three points are especially important in this regard. First, the change only applies to 29 CFR 1926.62. Second, the change only affects enforcement actions involving the Type-CE respirator used in abrasive-blasting that is manufactured by 3M as the Model 8100 Abrasive Blast Helmet. Third, the change is an interim one, pending a final determination by the Occupational Safety and Health Administration (OSHA) of the proper protection factor to be assigned to this class of respirators consequent to a forthcoming rulemaking on assigned protection factors (APF). Pending that determination, Regional Administrators shall ensure that the following policy is implemented. State designees are encouraged to adopt a similar enforcement policy for respiratory protection under 29 CFR 1926.62.

Based upon the 1987 Respirator Decision Logic developed by the National Institute of Occupational Safety and Health (NIOSH), OSHA in the Lead in Construction Standard designated an APF of 25 times the permissible exposure (PEL) for all Type-CE, continuous-flow, loose-fitting, atmosphere-supplying, airline abrasive-blast respirators (hood or helmet). With that assigned protection factor, this type of respirator would be acceptable for use only where airborne lead concentrations are less than or equal to 25 times the PEL of 50 ug/m(3), which is 1250 ug/m(3).

In a August 7, 1998, letter to OSHA, 3M Occupational Health and Safety Division provided documentation showing that their continuous flow Model 8100 Abrasive Blast Helmet achieved a considerably higher protection factor than the APF of 25 as currently assigned. 3M maintained that its respirator provides much greater protection and sought to have the APF under the Lead in Construction Standard (29 CFR 1926.62) elevated to 1000.

3M contracted with an acceptable independent third party to design, monitor, and interpret the results of a simulated workplace study of the Model 8100 under a test protocol that had been approved by OSHA. As a condition for granting relief, the Agency required that the results of the study demonstrate that the abrasive-blast respirator achieve, at a minimum, a protection factor rating of at least 20,000, and maintain positive pressure throughout the testing.

3M contracted with Los Alamos National Laboratory which designed, conducted, and interpreted the results of a study based on the OSHA-approved protocol. In the test, the

Model 8100 respirator achieved a minimum protection factor of greater than 20,000, and maintained positive pressure throughout the testing.

OSHA recognizes that a protection factor greater than 25 is appropriate for the 3M Abrasive Blast Helmet, Model 8100. The simulated workplace study carried out by the Los Alamos National Laboratory indicates that, if used properly, this respirator is acceptable for exposures to lead that are less than or equal to 1000 times the PEL (50,000 ug/m(3)). However, other data, and at least one field study, indicate that in the workplace these respirators may provide considerably less protection than indicated by the simulation study. Protection is degraded when the respirators are used in ways that do not conform to the manufacturer's specifications (e.g., the air supply hose is too long, the hose diameter is incorrect and/or the manufacturer's specified pressure is not maintained), or in ways that do not comply with the requirements of 29 CFR 1910.134 (e.g., the respirator is not inspected frequently enough for possible deterioration), which are incorporated by reference in the Lead in Construction Standard, 29 CFR 1926.62(f)(4).

Respirators will provide less protection than their design capability when used improperly. Examples of improper respirator use include the donning and doffing of respirators while employees are still in containment, or if they disconnect the air hose prior to leaving the exposure area. What is unusual in connection with these respirators is the extreme conditions under which they may be used in construction activities. Typically, abrasive blast respirators are used at very high levels of exposure and are subject to substantial and, at times, rapid deterioration due to damage caused by the high-speed, abrasive material used in the blasting. Also, at times, these respirators will sometimes be used near the limits of their protective capability. Consequently, workers using these respirators in abrasive- blasting operations may be subject to acute toxic exposures if the respirators do not perform properly. It is imperative, therefore, that these respirators be properly used.

For these reasons, OSHA will adopt the following approach in its enforcement policy with regard to these respirators:

First, this 3M model will be treated by OSHA as if it had an APF of 1000. Effective immediately for abrasive-blasting operations covered under the Interim Final Rule for Lead in Construction, the 3M Model 8100 Abrasive Blast Respirator is acceptable in abrasive-blasting atmospheres where the airborne level does not exceed 50,000 ug/m(3) (1000 times the PEL) of lead n air.

Second, OSHA will be very strict in assuring that this respirator is used only in accordance with the requirements of 29 CFR 1926.62. During compliance activities, CSHO's shall determine that all 3M respirators consist of the appropriate components, such as correctly sized air supply hoses and hose length, and that the required pressure range is maintained. If the respirator is not used in compliance with the manufacturer's specifications and with 29 CFR 1926.62, CSHO's will document the respiratory deficiencies. Violations related to documented deficiencies in the respirator will be cited.

Third, relief for this model applies only to lead in construction. The APF of 1000 does not apply to other air contaminants.

OSHA is now reviewing literature and comments for assigned protection factors for the inclusion in the revised respiratory protection standard 29 CFR 1910.134. OSHA will determine if an APF of 1000 should be assigned to the class of respirators addressed in this memorandum. The relief being granted to 3M, therefore, is interim relief only. Decisions in OSHA's final revised Respiratory Protection rule will supersede this decision and may reflect an APF different than 1000.

If you should have any further questions concerning this matter, please feel free to contact the Office of Health Compliance Assistance at (202) 693-2190.

Standard Interpretations - Table of Contents

Occupational Safety & Health Administration
200 Constitution Avenue, NW
Washington, DC 20210

XII. THE FEDERAL HAZARDOUS SUBSTANCES ACT

The Federal Hazardous Substances Act (FHSA) authorizes the Consumer Product Safety Commission (CPSC) to declare substances or mixtures of substances to be hazardous substances and to regulate, exempt or prohibit such substances under prescribed circumstances.[1] The CPSC adopted a regulation banning the shipment in interstate commerce after December 31, 1973 of any paint or other similar surface-coating material containing lead compounds with a lead content (calculated as the metal) exceeding 0.06% of the total weight of the contained solids or dried paint film. The CPSC also banned the shipment in interstate commerce after December 31, 1972 of any toy or other article bearing any paint or similar surface coating with such a lead content.[2]

The CPSC revoked the foregoing regulations when it adopted more comprehensive regulations under the Consumer Product Safety Act.[3] The CPSC's regulations under the Consumer Product Safety Act are discussed in the next Chapter.

---

[1] 15 U.S.C. § 1261 et seq.

[2] 16 CFR § 1500.17(a)(6)(ii)(A) and (B).

[3] 16 CFR § 1500.17(a)(6)(iii). See also 16 CFR § 1145.2.

XIII. THE CONSUMER PRODUCT SAFETY ACT

A. General

The Consumer Product Safety Act (CPSA) authorizes the Consumer Product Safety Commission (CPSC) to regulate consumer products. Among other powers, the CPSC is authorized to promulgate a rule banning a consumer product whenever it finds that the product presents an unreasonable risk of injury and no feasible standard adopted under the CPSA would adequately protect the public from such unreasonable risk of injury.[1]

B. Prohibited Consumer Products (Lead-Containing Paint)

The CPSC has banned the following consumer products manufactured after February 27, 1978, unless exempted: (1) paint and similar surface coating materials which are lead-containing paint, (2) toys and other articles intended for use by children that bear lead-containing paint, and (3) furniture articles that bear lead-containing paint. The CPSC defined lead-containing paint as paint or similar surface coating materials containing lead or lead compounds in which the lead content (calculated as lead metal) exceeds 0.06% by weight of the total nonvolatile content of the paint or the weight of the dried paint film.[2]

In effect, the ban prevented the use of lead-based pigments and also lead-based driers in consumer paints and coatings, toys and other articles intended for use by children, and furniture articles.

C. Exemptions and Labelling

The CPSC's regulations at 40 CFR § 1303.3 exempt the following coatings, provided that they comply with labeling requirements:

1. Agricultural and industrial equipment refinish coatings.

2. Industrial (and commercial) building and equipment maintenance coatings, including traffic and safety marketing coatings.

3. Graphic art coatings (i.e., products marketed solely for application on billboards, road signs, and similar uses and for identification marking in industrial buildings).

---

[1] 15 U.S.C. § 2057. CPSC also administers the Federal Hazardous Substances Act.

[2] 16 CFR Part 1303; see also 16 CFR §1145.2

4. Touchup coatings for agricultural equipment, lawn and garden equipment, and appliances.

5. Catalyzed coatings marketed solely for use on radio-controlled model powered aircraft.

The exempted coatings must bear on the main panel of their label the signal "Warning" and the statement: "Contains Lead. Dried Film Of This Paint May Be Harmful If Eaten Or Chewed." The label must also contain the following statement or its practical equivalent:

> "Do not apply on toys and other children's articles, furniture, or interior surfaces of any dwelling or facility which may be occupied or used by children.
>
> Do not apply on exterior surfaces of dwelling units, such as window sills, porches, stairs or railings, to which children may be commonly exposed.
>
> Keep out of reach of children."

If the foregoing statement is not on the main panel, a statement must be made on the main panel as follows: "See other cautions on the side (reverse) panel." The placement, conspicuousness and contrast of the label statements must comply with the requirements of the Federal Hazardous Substances Act contained in the CPSC's regulations at 16 CFR § 1500.121.

The CPSC's regulations exempt the following consumer products without imposing a labelling requirement:

1. Mirrors which are part of furniture articles to the extent that they bear lead-containing backing paint.

2. Artists' paints and related materials.

3. Metal furniture articles (but not metal children's furniture) bearing factory applied (lead) coatings.

D. 1992-1993 Regulatory Investigation

In 1992, the CPSC announced a regulatory investigation to determine whether it should revise its regulations to lower the permissible level of lead content in paint to 0.01%.[3] In 1993, the CPSC terminated this regulatory investigation because the information obtained showed only a small risk of adverse health effects from the lead content of currently marketed paints and because there was insufficient information showing that the benefits of a further reduction would bear a reasonable relationship to the costs. Nationwide sampling by the CPSC showed that 90% of the samples contained less than 0.01% lead and 96% contained less than 0.02% lead. The average lead content was 0.004%.[4]

E. Other Prohibited Products (Lead-Containing Candlewicks)

Among the other products prohibited by the CPSC are metal-cored candlewicks with a lead content over 0.06% of the total weight of the metal core. Candles made with such wicks are also prohibited. Candles and candlewicks which meet the lead threshold must also meet packaging and labelling requirements.[5]

F. Enforcement Programs and Guidance

The CPSC maintains an active enforcement program to protect against consumer products containing lead, especially products imported from other nations, because there are nations where they may lawfully be sold. The CPSC has provided guidance on lead-containing and other products such as vinyl miniblinds, crayons, figurines used as gamepieces, and children's jewelry.[6]

Being specially concerned about recalls of children's metal jewelry, the CPSC published an interim enforcement policy dated February 3, 2005. The policy explained that young children should not chronically ingest more than 15 micrograms of lead per day from consumer products in order to avoid exceeding the scientifically recognized threshold of concern

[3] 57 Fed. Reg. 18418 (April 30, 1992).

[4] 8 Fed. Reg. 63311 (Dec. 1, 1993).

[5] 16 CFR §1500.17(a)(13)(i)(A).

[6] 16 CFR §1500.230.

of 10 micrograms per deciliter of blood. The CPSC warned that children who wear metal jewelry containing accessible lead may ingest the lead by hand to mouth activity and even by swallowing all or part of the jewelry. The CPSC staff recommended that children not ingest more than 15 micrograms of accessible lead in a short period. The CPSC attached a standard operating procedure (test protocol) for determining the lead and its availability in children's metal jewelry including a screening test and an acid extraction test. In November 2005, the CPSC issued an FAQ answering questions about the methods including the acceptability of ASTM methods using inductively coupled plasma atomic emission spectrometry (ICP-AES), flame atomic absorption spectrometry (FAAS), graphite furnace atomic absorption spectrometry (GFAAS), and inductively coupled plasma mass spectrometry (ICP-MS) equipment. The policy, the standard operating procedure and the FAQ are available on the CPSC's website.

Reinforcing the CPSC's concerns, Reebok International Ltd. voluntarily cooperated with the CPSC in a recall in early 2006 of heart-shaped charm bracelets manufactured in China following a report that a 4 year old child in Minneapolis had died of lead poisoning after swallowing a piece from one of the bracelets. The CPSC's announcement dated March 3, 2006 appears on its website, and further information appears on the website of the Centers for Disease Control.

The CPSC reportedly voted on December 11, 2006 to instruct its staff to prepare an ANPR to declare children's jewelry containing over 0.06% lead by weight in its metal components to be a hazardous substance. The ANPR was published in the Federal Register, Vol. 72, page 920 on January 1, 2007.

An informative and well-written article on these and related subjects titled "Lead in Consumer Products an Ongoing Concern for Regulators and Litigators" recently appeared in the Toxics Law Reporter.[7]

---

[7] 21 Toxics Law Reporter (BNA) 612, Laurenza, June 29, 2006.

XIV. THE FEDERAL FOOD, DRUG AND COSMETICS ACT

A. General

The Federal Food, Drug and Cosmetics Act (FFDCA) was adopted to protect the public from adulterated or misbranded food, drugs and cosmetics.[1] The FFDCA is administered and enforced by the Food and Drug Administration (FDA) of the U.S. Department of Health and Human Services. The FDA shares regulation on these fields with other government agencies such as the U.S. Department of Agriculture (USDA), which regulates meats, and the EPA, which regulates insecticides, fungicides and rodenticides.

The FDA has adopted extensive and detailed regulations under the FFDCA. They regulate food and also food additives. Regulation of food additives was required in order to authorize the FDA to protect food from adulteration.

Food additives include not only substances intentionally added to food, but any other substances that may reasonably be expected to become a component or affect the characteristics of food when used as intended. Thus, food additives include substances used in manufacturing, processing, storing, packaging or transporting food, whether for humans or animals. For example, food additives include colorants, pesticide residues and food contact surfaces in packaging materials.

The FDA's regulations include a wide variety of subjects. Examples of the FDA's standards and other rules for food are as follows:

- Food quality
- Food additives, direct and indirect
- Generally recognized as safe (GRAS) substances
- Color additives
- Prohibited substances
- Labelling
- Exemptions and tolerances
- Good laboratory practices
- Good manufacturing practices

---

[1] 21 U.S.C. § 301 et seq.

- Petitions
- Inspections, recordkeeping and reports

The FDA adopts standards which it develops directly. The FDA also reviews all standards adopted by the Codex Alimentarius Commission of the Joint Food and Agriculture Commission of the United Nations and the World Health Organization. The Codex Alimentarius includes food and food additives as well as labeling and methods for sampling and testing. The Codex standards are not legally enforceable unless and until individual nations adopt them as laws or regulations. The FDA has undertaken to review all Codex food standards, but has reserved freedom to accept them with changes and not to accept them.[2]

The cooperation between the EPA and the Food Codex Commission of the National Academy of Sciences is described later in this Chapter.

B. Regulations Relating to Lead in Food

Lead is not one of the metals naturally contained in the human body and is not a desired ingredient of food. However, lead has for centuries become a food ingredient because of its presence in food additives, direct and indirect. Examples of food contact materials that may add lead to food are containers, container coatings and solder, water piping and solder, food machinery and equipment, and dust which reaches food directly or indirectly through the food chain.

Lead is also extensively used in drugs and cosmetics, and a number of folk remedies contain lead. Lead may become an ingredient of drugs and cosmetics as direct and indirect additives.

Examples of the FDA's standards for lead and lead compounds in food and additives are provided in the following paragraphs:

Food Additives for Human Food. The FDA's regulations on food additives, including indirect food additives, contain limits on lead content for many additives.[3] For example, the limit on lead in glycerides and polyglycides of hydrogenated vegetable oils is 0.1

---

[2] 21 CFR § 130.6.

[3] 21 CFR §172.736, §172.859 and §172.869.

mg/kg, for sucrose fatty acid esters is 10 ppm, and for sucrose oligoesters is 1.0 mg/kg. The list of specific substances affirmed as generally recognized as safe (GRAS) also contains limits on lead content. For example, the GRAS limit for aconitic acid for heavy metals (as Pb) is 10 ppm and for nisin preparation is 2 ppm. For many additives such as adipic acid and benzoic acid, the GRAS regulations require that the ingredient meet the specifications in the Food Chemicals Codex.

In 1995, the FDA amended its food additive regulations to prohibit the use of lead solder to manufacture cans for packaging foods by adopting a rule that food packaged in any container using lead in can solder shall be deemed adulterated.[4] The ban became effective six months after June 27, 1995 in order to allow for sale of existing stocks of lead-soldered canned foods. The FDA commented in its discussion of the rule that food manufacturers in the United States had voluntarily been discontinuing use of lead solder in cans for packaging food. Thus, an important effect of the rule was to restrict the use of lead solder by foreign food manufacturers who export canned food to the United States.

In 1996, the FDA amended its food additive regulations to prohibit capping of wine bottles with tin coated lead foil capsules.[5]

Lead in Ceramicware and Silver-Plated Hollowware. In a pamphlet titled "The Food Defect Action Levels," the FDA established the following levels of lead content at which it will take enforcement action against a product:

Ceramicware	
Flatware	3.0 ug/ml
Small hollowware	2.0 ug/ml
Large hollowware	1.0 ug/ml
Cups and mugs	0.5 ug/ml
Pitchers	0.5 ug/ml
Silver-plated hollowware	

[4] 21 CFR § 189.240; 60 Fed. Reg. 33106 (June 27, 1995).

[5] 21 CFR §189.301.

Product for adult use	7.0 ug/ml
Product for children's use	0.5 ug/ml

The FDA also requires that decorative ceramicware containing lead be labeled to warn that it is not intended for food use.[6]

Bottled Water. The FDA maintained a standard for lead in bottled water. Effective November 21, 1994, the limit was reduced to 0.005 mg/L.[7]

Color Additives for Food. The FDA has established limits for the lead content in a number of color additives which may be used in food without certification by the FDA. Examples of the limits are dehydrated beets (10 ppm), ultramarine blue (10 ppm), canthaxanthin (10 ppm), caramel (10 ppm); B-Apo-8-carotenal (10 ppm), toasted partially defatted cooked cottonseed flour (10 ppm), grape color and grape skin extracts (10 ppm), synthetic iron oxide for human food (10 ppm), synthetic iron oxide for dog and cat food (20 ppm), and titanium dioxide (10 ppm).[8]

Color Additives for Drugs. The FDA has established limits for the lead content in a number of color additives which may be used in drugs without certification by the FDA. Examples of the limits are alumina (10 ppm), chromium-cobalt-aluminum oxide (30 ppm), ferric ammonium citrate (20 ppm), potassium sodium copper chlorophyllin (20 ppm), mica-based pearlescent pigments (4 ppm), dihydroxyacetone (20 ppm), bismuth oxychloride (20 ppm), synthetic iron oxide (10 ppm), ferric ammonium ferrocyanide (20 ppm), ferric ferrocyanide (20 ppm), chromium hydroxide green (20 ppm), chromium oxide green (20 ppm), guanine (20 ppm), pyrogallol (20 ppm), pyrophyllite (20 ppm), logwood extract (70 ppm), mica (20 ppm), talc (20 ppm), aluminum powder (20 ppm), bronze powder (20 ppm), copper powder (20 ppm) and zinc oxide (20 ppm).[9]

---

[6] 21 CFR §109.15; 59 Fed Reg. 1638 (January 12, 1994).

[7] 21 CFR § 165.110.

[8] See 21 CFR Part 73, Subpart A, §§ 73.1 to 73.615.

[9] 21 CFR Part 73, Subpart B, § 73.1001 et seq.

Color Additives for Cosmetics. The FDA has established limits for the lead content in a number of color additives which may be used in cosmetics without certification by the FDA. Examples of the limits are bismuth citrate (20 ppm), disodium EDTA-copper (20 ppm), guiazulene (20 ppm), henna (20 ppm), iron oxides (10 ppm), lead acetate (0.6% weight to volume in the cosmetic), silver (10 ppm), ultramarines (20 ppm) and manganese violet (20 ppm).[10]

Color Additives Which Must Be Certified. The FDA has established limits for the lead content in a number of color additives, subject to certification by the FDA: For example, the limits for food colorants from FD&C Blue No. 1 to FD&C Yellow No. 6 are 10 ppm.[11]

Food Additives for Animal Feed and Drinking Water: The FDA's regulations set limits for lead content in food additives for animal feed and drinking water. For example, the limit for diammonium phosphate is 30 ppm for heavy metals, as lead, and for diatomaceous earth is 15 ppm.[12]

C. FDA Proposal to Adopt Stricter Lead Standards.

In 1994, the FDA published an advance notice of proposed rulemaking announcing its intent to propose new lower specifications for the most heavily used food and color additives and GRAS ingredients. The FDA said that it had for many years considered that the public was adequately protected by specifications of 10 ppm for lead and 40 ppm for total heavy metals (as lead). However, the Food Chemicals Codex Committee of the National Academy of Sciences began lowering specifications for high volume sweeteners during the 1980s to 1 ppm, then to 0.5 ppm, and then to 0.1 ppm in voluntary programs arranged with industry.

---

[10] 21 CFR Part 73, Subpart C, § 73.2030 et seq.

[11] 21 CFR Part 74, Subpart A, § 74.101 et seq.

[12] 21 CFR §573.320 and §573.340.

D. The Food Chemicals Codex

The Food Chemicals Codex, 5th Ed., January 1, 2004 is published by the Food and Nutrition Board of the Institute of Medicine of the National Academies. The Codex specifies a lead content limit of 0.1 ppm for canola oil, castor oil, dextrose, fructose and glucose syrups, and 0.5 ppm in refined products such as citric acid, maltodextrin and polydextrose. However, the Codex allows higher lead content in other food chemicals such as caramel (2 ppm), carnauba wax (5 ppm), aluminum sulfate (3 ppm), copper sulfate (4 ppm) and diatomaceous earth (10 ppm). Appendix IIIB of the Codex prescribes tests for lead content. Unless otherwise specified, the Dithizone method is to be used. However, the graphite furnace atomic absorption spectrophotometric (GFAAS) method is prescribed for analysis of substances containing less than 1 ppm of lead.

E. Cosmetics Tolerances

The FDA has not established general tolerances for lead in cosmetics, except for coal tar colors. The FDA stated in 1940 that it did not object to use of titanium dioxide, a white pigment, in cosmetics if it contains less than 20 ppm of lead. Such a level in the colorant would result in a lower level in the finished cosmetic product.

F. Candy Likely to Be Consumed by Small Children

In November 2006, the FDA published a Guidance for Industry titled "Lead in Candy Likely to be Consumed Frequently by Small Children: Recommended Maximum Level and Enforcement Policy." The Guidance provides a recommended maximum lead level of 0.1 ppm. The FDA considers that level to be achievable with use of good manufacturing practices and to be protective of human health. See 71 Fed. Reg. 67881, Nov. 24, 2006.

G. Folk Remedies and Cosmetics

The FDA issues import alerts and other warnings about lead in folk remedies and cosmetics. For example, the FDA issued an automatic detention alert on September 22, 1992 for Mexican folk remedies containing lead: Alarcon, Azarcon, Coral, Greta, Liga, Maria Luisa and Rueda. The FDA has also warned about lead in herbal medicines and unapproved dyes in eye cosmetics such as kajal, surma and kohl from far and middle eastern countries. In its Draft

Toxicological Profile for Lead, the ASTDR also warned about such folk remedies as well as Bali Goli, Ghasard, Koo Sar and Pay-loo-ah.

XV. HAZARDOUS MATERIALS TRANSPORTATION ACT

A. General

The Hazardous Material Transportation Act (HMTA)[1] requires that shippers, transporters and persons who manufacture, sell or perform services related to packages or containers of hazardous, radioactive or explosive materials in excess of specified quantities must submit a registration statement on Form 5800.2 to the Pipeline and Hazardous Materials Safety Administration (PHSMA) of the U.S. Department of Transportation (DOT) which has administered the HMTA for many years. Although little known to the general public, the programs and experience of the DOT and the PHSMA, formerly called the Research and Special Programs Administration, have long provided expert and important protection of public safety.

Hazardous materials are substances and materials in quantities or forms that pose an unreasonable risk to health and safety or property when transported in commerce. They include, without limitation, explosives, radioactive materials, etiological agents, flammable liquids or solids, poisons, oxidizing or corrosive materials, and compressed gases.

Hazardous materials include hazardous waste.[2] The definition of hazardous waste is any waste subject to the Hazardous Waste Manifest Requirements of the U.S. Environmental Protection Agency specified in 40 CFR Part 262.[3]

B. PHSMA Regulations

Under the HMTA and earlier laws, the PHSMA has adopted regulations creating a comprehensive system for the safe transportation of hazardous materials by rail, aircraft, vessel and public highway.[4] The regulations prescribe test methods to determine hazards. They require warning labels on containers and packages and warning placards on vehicles. They prescribe specifications for containers and packages and safe loading and unloading procedures. They require accident reports and remedial action as well as more routine record keeping and reporting. The DOT conducts inspections to monitor compliance.

---

[1] 49 U.S.C. § 5101 et seq. (formerly § 1801 et seq.)

[2] 49 CFR §§ 171.3, 71.8 and 172.101.

[3] See 40 CFR §262.11 on hazardous waste determination.

[4] 49 CFR Parts 171-180.

PHSMA's regulations are enforced by DOT administrations for shipments by the rail, air, and motor vehicle carriers which they regulate, i.e., the Federal Railway Administration (FRA), the Federal Aviation Administration (FAA), and the Federal Motor Carrier Safety Administration (FMCSA), and for shipments by water by the U.S. Coast Guard which is part of the Department of Homeland Security (DHS). Each administration has adopted implementing regulations. PHSMA regulates intermodal shipments.

C. Lead Compounds Listed in the Hazardous Materials Table

A number of lead compounds are listed in the Hazardous Materials Table which follows 49 CFR § 172.101. The Table describes for each compound the applicable hazard class or division; the United Nations or North American identification number; the packaging group; the label; any special provisions; the packaging authorizations for nonbulk and bulk packaging; quantity limitations for aircraft or rail shipment; and vessel stowage requirements. The following are the listed lead compounds, the identification numbers and the required levels:

Lead acetate	UN1616	KEEP AWAY FROM FOOD
Lead arsenates	UN1617	POISON
Lead arsenites	UN1618	POISON
Lead azide (dry)	- - - - -Forbidden- - - - -	
Lead azide (wet)	UN0129	EXPLOSIVE 1.1A
Lead compounds, soluble, n.o.s.	UN2291	KEEP AWAY FROM FOOD
Lead cyanide	UN1620	POISON
Lead dioxide	UN1872	OXIDIZER
Lead dross	- - - - -See Lead sulfate- - - -	
Lead nitrate	UN1469	OXIDIZER, POISON
Lead nitro-resorcinate (dry)	- - - - -Forbidden- - - - -	
Lead perchlorate, solid	UN1470	OXIDIZER, POISON
Lead perchlorate, solution	UN3408	

Lead peroxide	- - - - -See Lead dioxide- - - -	
Lead phosphite, dibasic	UN2989	FLAMMABLE SOLID
Lead picrate (dry)	- - - - -Forbidden- - - - -	
Lead styphnate (dry)	- - - - -Forbidden- - - - -	
Lead styphnate (wet)	UN0130	EXPLOSIVE 1.1A
Lead sulfate, >3% free acid	UN1794	EXPLOSIVE 1.1A
Lead trinitroresorcinate (wet) - No less than 20% water or mixture of alcohol and water, by mass	UN0130	EXPLOSIVE 1.1A

D. Reportable Quantities for Spills and Releases

Appendix A, Table 1 to the Hazardous Materials Table lists the reportable quantity (RQ) of lead and several hazardous lead compounds that must be reported to the EPA's CERCLA hotline in the event of a spill or other release.[5] The substances are as follows: Lead, lead acetate, lead arsenate, lead bis (acetato-0) tetrahydrate, lead chloride, lead fluoroborate, lead fluoride, lead iodide, lead nitrate, lead phosphate, lead stearate, lead subacetate, lead sulfate, lead sulfide and lead thiocyanate. The RQ for all of them is 10 pounds (4.54 kilograms), except lead arsenate for which the RQ is 1 pound (0.454 kilograms).

Appendix A, Table 1 also lists the RQs for the hazardous wastes listed under RCRA including wastes listed because they contain lead constituents. See Chapter IX for a description of these wastes. The RQ for these wastes is generally 10 pounds (4.54 kilograms).

[5] 49 CFR § 172.101, App. A

Appendix A, Table 2 lists the RQs for radionuclides including the radioactive lead isotopes 195m through 203, 205, 209 through 212, and 214. The RQs range from 0.01 Curies to 1,000 Curies as follows:

Isotope	Ci (TBq)
Lead-195m	1000 (37)
Lead-198	100 (3.7)
Lead-199	100 (3.7)
Lead-200	100 (3.7)
Lead-201	100 (3.7)
Lead-202	1 (.037)
Lead-202m	10 (.37)
Lead-203	100 (3.7)
Lead-205	100 (3.7)
Lead-209	1000 (37)
Lead-210	0.01 (.00037)
Lead-211	100 (3.7)
Lead-212	10 (.37)
Lead-214	100 (3.7)

E. List of Marine Pollutants

Several lead compounds appear on the list of marine pollutants contained in 40 CFR § 172.101, Appendix B. The listed lead compounds are as follows:

Lead acetate

Lead arsenates

Lead arsenites

Lead compounds, soluble, n.o.s.

Lead cyanide

Lead nitrate

Lead perchlorate, solid or solution

Lead tetraethyl

Lead tetramethyl

Nonbulk packaging containing these compounds must be marked with a triangular symbol containing a drawing of a fish overlaid with an "X" and the words "Marine Pollutant." The symbol, letters and border must be black and the background must be of contrasting color to the surface to which the work is affixed.[6]

---

[6] 49 CFR § 172.322.

XVI. THE LEAD-BASED PAINT POISONING PREVENTION ACT, 42 U.S.C. § 4801, AND THE RESIDENTIAL LEAD-BASED PAINT HAZARD REDUCTION ACT, 42 U.S.C § 4851

A. General

Major Federal programs relating to lead-based paint originated with The Lead-Based Paint Poisoning Prevention Act of 1971 (the "Poisoning Prevention Act"). 42 U.S.C. § 4801 et seq. The programs were extensively supplemented by the Residential Lead-Based Paint Hazard Reduction Act of 1992 (the "Residential Hazard Reduction Act").[1]

The two Acts require the U.S. Department of Housing and Urban Development (HUD) to eliminate insofar as practicable the hazard of lead-based paint poisoning in federally financed and subsidized housing. The two Acts also established programs administered by several other federal government agencies and include amendments to other federal laws. For example, the CPSC's ban in 1978 of the sale of lead-based paint for consumer use was based on 1976 amendments to the Poisoning Prevention Act. The EPA's regulations on lead-based paint activities and OSHA's construction industry standard for workplace exposure were based on amendments to TSCA contained in the Residential Hazard Reduction Act.

This Chapter summarizes the two Acts and the regulations adopted by HUD and the other federal housing agencies to implement them. The programs administered by agencies other than HUD and the General Services Administration (GSA) are discussed in the Chapters on the laws which those agencies administer.

B. The Lead-Based Paint Poisoning Prevention Act of 1971

In 1971, the Poisoning Prevention Act was adopted to require reduced levels of lead in paint in federally financed and subsidized housing and to fund screening and research programs. In 1973, it was amended to require that the U.S. Department of Housing and Urban Development (HUD) take steps to eliminate insofar as practicable the hazards of lead-based paint poisoning in federally financed and subsidized housing.[2]

---

[1] 42 U.S.C. § 4851 et seq.

[2] 42 U.S.C. § 4801 et seq.

C. HUD and GSA Regulations and Programs After 1971

During the years since 1971, HUD has adopted numerous regulations which now appear primarily in 24 CFR Part 35, but also in other parts of its regulations. HUD's regulations are summarized later in this Chapter.

The GSA also adopted implementing regulations. For example, the GSA's regulations on utilization and disposal of hazardous materials and certain categories of property repeat the CPSC's ban of lead-based paint for consumer use as well as on children's toys and furniture articles bearing lead-based paint. The regulations require that excess banned products be destroyed, but allow stripping and refinishing with a nonhazardous coating in accordance with OSHA regulations. The regulations contain exemptions which require warning labels. They also contain utilization, donation and sales requirements for exempt products that require warning statements. Abandonment in any manner allowing acquisition and use is prohibited. Disposal must follow prescribed procedures.[3]

As the years progressed, HUD developed major programs related to lead-based paint and its detection and control or abatement. Through its programs and funding, HUD sponsored the growth of a significant industry. HUD provided grants for studies and for rehabilitation of federally financed and subsidized housing under several laws including the National Housing Act (42 U.S.C. § 1441 et seq.) and the Housing and Community Development Act (42 U.S.C. § 5301 et seq.). These laws have perennially been amended and expanded by legislation such as the McKinney Homeless Assistance Amendments Act of 1988, the Cranston-Gonzalez National Affordable Housing Act of 1990, and the Housing and Community Development Act Amendments of 1992. Among the grant programs are Section 8 Grants, Community Development Block Grants and HOME Grants.

Grants from HUD, combined with the research funded by the Department of Health and Human Services (DHHS), have provided hundreds of millions of dollars for inspection, risk assessment, abatement, research, consultation, legal services and advocacy

---

[3] 41 CFR § 101-42.1102-7.

activities relating to lead-based paint. In April, 1990, HUD published Interim Guidelines for hazard identification and abatement of lead-based paint in Public and Indian Housing Programs assisted under the Comprehensive Improvement Assistance Program (CIAP), 55 Fed. Reg. 14556 (April 18, 1990). The Interim Guidelines contained extensive requirements and also valuable information about methods and technologies. Among the guidelines were testing methods and equipment, qualifications and training of abatement contractors and workers, abatement choices and methods, worker protection, cleanup and disposal. The appendices included descriptions of state and local laws and extensive additional data.

In 1990, HUD presented a report to Congress titled a Comprehensive and Workable Plan for the Abatement of Lead-Based Paint in Privately Owned Housing in response to 1987 amendments to the Poison Prevention Act. In 1991, HUD issued a report titled "The HUD Lead-Based Paint Abatement Demonstration (FHA), August, 1991," describing in detail the multi-city abatement projects which it had sponsored. These reports contained extensive information for those interested in the hazards and abatement of lead-based paint.

D. <u>The Residential Lead-Based Paint Hazard Reduction Act of 1992</u>

The Residential Hazard Reduction Act begins with Congressional findings that "low-level lead poisoning" is widespread among millions of children (with minority and low-income communities disproportionately affected) and causes IQ deficiencies, reading and learning disabilities, impaired hearing, reduced attention span, hyperactivity, and behavior problems. See Chapter XXI for the effect of these findings on lead litigation.

1. <u>Amendments to the Poisoning Prevention Act</u>.

The Residential Hazard Reduction Act amended the Poisoning Prevention Act to require HUD to establish procedures to eliminate as far as practicable the hazards of lead-based poisoning in housing covered by an application for mortgage insurance in housing assistance payments under any HUD program. The procedures must apply to all target housing (i.e., housing constructed before 1978) and must provide for measures to conduct risk assessments, inspections, interim controls and abatement of lead-based paint hazards. The procedures must, at minimum, include (A) provision of lead hazard information pamphlets; (B) periodic risk assessments and interim controls according to a statutory schedule from January 1, 1996 to January 1, 2002; (C) inspection for lead-based paint before federally funded renovation or rehabilitation likely to disturb painted surfaces; (D) reduction of lead-based paint hazards in the course of rehabilitation projects receiving less than $25,000 per unit in federal funds; (E) abatement of lead-based paint hazards in the course of substantial rehabilitation projects receiving more than $25,000 per unit in federal funds; and (F) notice to occupants describing any risk assessment, inspection and reduction activities undertaken including the locations of any remaining lead-based paint on a surface-by-surface basis.

Beginning January 1, 1995, HUD was required to inspect and abate lead-based paint hazards in all federally owned target housing constructed prior to 1978. Beginning January 1, 1995, HUD was required to inspect all federally owned target housing constructed between 1960 and 1978 for lead-based paint and lead-based paint hazards and make the results available to prospective purchasers. However, HUD was authorized to waive the inspection requirements for 1960-1978 housing in which a federally funded risk assessment, performed by a certified contractor, has determined no lead-based paint hazards are present.

2. <u>Amendments to the OSH Act</u>.

The Residential Hazard Reduction Act required that the Secretary of Labor issue an interim final lead standard pursuant to the OSH Act covering the construction industry that would provide protection to construction workers from occupational exposure to lead as effective as the HUD guidelines and OSHA's lead standard for general industry. Among other things, the amendments required that OSHA establish a PEL of 50 ug/m^3, 8 hour TWA and an "action level" of ug/m^3 so that the construction industry standards would correspond to those applicable to general industry. OSHA complied by publishing an interim final rule.[4] The rule adopted by OSHA is described in Chapter XII.

[4] 58 FR 26627, May 4, 1993.

3. Amendments to the Toxic Substances Control Act

Title X of the Residential Hazard Reduction Act amended TSCA to require the EPA, in consultation with OSHA, HUD and the National Institute for Occupational Safety and Health (NIOSH), to issue regulations governing lead-based paint activities to ensure that individuals engaged in such activities are properly trained; that training programs are accredited; and that persons engaged in such activities are certified. The regulations must require that risk assessment, inspection, and abatement activities performed in "target housing" (i.e., all housing constructed prior to 1978, with certain exceptions) must be performed by certified contractors. The amendments defined lead-based paint as any paint or other surface coating containing in excess of 1.0 mg/cm^2 or 0.5% by weight or such lower levels in target housing as may be established by HUD or for other coatings by the EPA. Among other things, the amendments required the EPA to adopt regulations to govern renovation and remodelling activities in target housing that may create lead exposure risk and to identify lead-based paint hazards, lead contaminated dust, and lead-contaminated soil. They also required the EPA to conduct a program to promote safe, effective and affordable monitoring, detection and abatement of lead-based and related hazards.

The amendments to TSCA and the EPA's existing and proposed actions to implement them are described in greater detail in Chapter XVII.

4. Other Provisions of the Residential Hazard Reduction Act

HUD was authorized to provide grants to state and local governments to evaluate and reduce lead-based paint hazards in priority housing that is not federally assisted housing, federally owned housing, or public housing.[5]

HUD was directed to establish a Task Force to make recommendations on expanding resources and efforts to evaluate and reduce lead-based paint hazards. The Task

[5] 42 U.S.C. § 4852.

Force must include representatives of government, labor, industry, landlords, tenants, lenders, insurers, public housing agencies, housing and health advocacy organizations and others.[6]

HUD was required to issue guidelines for the conduct of federally supported work involving risk assessments, inspections, interim controls and abatement of lead-based paint hazards. The guidelines must be based on criteria that measure the condition of the housing (and the presence of children under age 6 for the purpose of risk assessments) and not upon criteria that measure the health of the residents of the housing.[7] HUD announced the availability of its "Guidelines for the Evaluation and Control of Lead-Based Paint Hazards in Housing" on August 25, 1995.[8] The Guidelines can be obtained from HUD's website.

HUD and the EPA were required to issue regulations under TSCA to become effective October 28, 1995 requiring that every seller and lessor of "target housing" meet several requirements for the benefit of the buyer or tenant. "Target housing" means any housing constructed prior to 1978 except certain housing for the elderly and disabled. The requirements are to furnish a lead hazard information pamphlet prepared by the EPA, disclose the presence of any known lead-based paint or hazards in the housing, and permit a 10-day period to conduct a risk assessment or inspection. Contracts for purchase or sale of any interest in target housing must contain a Lead Warning Statement and a statement signed by the purchaser that evidences compliance with the requirements of the regulations. Treble damages, attorneys fees and civil penalties are provided for violations.[9] The regulations adopted by HUD are described in this Chapter and the regulations adopted and proposed by the EPA are described in Chapter XVII.

HUD, in cooperation with other federal agencies, is required to conduct research on strategies to reduce the risk of lead from sources other than lead-based paint including

---

[6] 42 U.S.C. § 4852a.

[7] 42 U.S.C. § 4852c.

[8] 60 Fed. Reg. 44562 (August 25, 1995).

[9] 42 U.S.C. § 4852d.

exterior soil and interior lead dust in carpets, furniture and forced air ducts.[10] HUD must also conduct research to:

(1) develop improved methods for evaluating lead-based paint hazards in housing;

(2) develop improved methods for reducing lead-based paint hazards in housing;

(3) develop improved methods for measuring lead in paint films, dust, and soil samples;

(4) establish performance standards for various detection methods, including spot test kits;

(5) establish performance standards for lead-based paint hazard reduction methods, including the use of encapsulants;

(6) establish appropriate cleanup standards;

(7) evaluate the efficacy of interim controls in various hazard situations;

(8) evaluate the relative performance of various abatement techniques;

(9) evaluate the long-term cost-effectiveness of interim control and abatement strategies; and

(10) assess the effectiveness of hazard evaluation and reduction activities funded by this chapter.

The Residential Hazard Reduction Act includes the EPA in the major funding programs previously handled by HUD and DHHS. For example, the EPA published a notice of funds availability ($11,200,000) for grants to develop and carry out state accreditation and certification programs for lead-based paint professionals.[11] The EPA published a notice of request for proposals to provide training for lead-based paint abatement workers without mentioning the amount available.[12] The EPA published a notice of funds availability

---

[10] 42 U.S.C. § 4854.

[11] 59 Fed. Reg. 10131 (March 3, 1994).

[12] 59 Fed. Reg. 18811 (April 20, 1994).

($12,500,000) for grants to carry out authorized state accreditation and certification programs for lead-based paint professionals.[13]

E. Department of Housing and Urban Development Regulations[14]

HUD has adopted numerous regulations applicable to the various programs which it administers:

1. Disclosures Upon Sale or Lease of Residential Property

HUD has regulations requiring disclosure of known lead-based paint hazards upon sale or lease of target housing, i.e., housing constructed before 1978.[15] They correspond to the EPA's regulations under the Toxic Substances Control Act described in detail in Chapter XVII. Like the EPA's regulations, HUD's disclosure regulations require sellers and lessors to provide the EPA's pamphlet titled "Protect Your Family from Lead in Your Home" to purchasers and lessees of target housing. Sellers and lessors must also disclose any known lead-based paint and/or lead-based paint hazards and furnish available information about them including records and reports. Like the EPA's regulations, HUD's disclosure regulations also require that sellers permit purchasers to have a period of time for a risk assessment or inspection unless the opportunity is waived in writing. Contracts for sale or lease must contain a Lead Warning Statement and other required information including a certification and acknowledgement of receipt of disclosures. Sellers and lessors must retain copies of the certification and acknowledgement information for three years. Agents for sellers and lessors are required to assure compliance with the regulations and must also retain certification and acknowledgement information for three years.

---

[13] 60 Fed. Reg. 10583 (February 27, 1995).

[14] 24 CFR Part 35.

[15] 24 CFR §35.80 et seq.

2. General Requirements and Definitions for All Programs

HUD has general lead-based paint requirements and definitions for programs involving target housing that is federally owned or receiving federal assistance.[16]

Summaries of selected key definitions are as follows:

"Abatement" means any set of measures designed to permanently eliminate lead-based paint or lead-based paint hazards. The definition includes examples such as removal, permanent enclosure or encapsulation, and replacement.

"Chewable surface" means an interior or exterior surface painted with lead-based paint that a young child can mouth or chew. It does not include hard metal substrates and other materials that cannot be dented by the bite of a young child.

"Dust lead hazard" means surface dust that contains a dust-lead loading (area concentration of lead) equal to or exceeding the levels promulgated by the EPA at 40 CFR §745.65.

"Encapsulation" means the application of a covering or coating that acts as a barrier between lead-based paint and the environment and that relies for its durability on adhesion between the encapsulation and the painted surface and on the integrity of the existing bonds between paint layers and between the paint and the substrate. The definition adds that encapsulation may be used as an abatement method if designed and performed to be permanent.

"Enclosure" means use of rigid, durable construction materials mechanically fastened to the substrate to act as a barrier between lead-based paint and the environment. The definition adds that enclosure may be used as an abatement method if designed to be permanent.

"Environmental intervention blood lead level" means a confirmed concentration of lead in whole blood equal to or greater than 20 ug/dL for a single test or 15-19 ug/dL in two tests taken at least 3 months apart.

"Evaluation" means a risk assessment, lead hazard screen, lead-based paint inspection, paint testing, or a combination of them to determine the presence of lead-based hazards or lead-based paint.

---

16 24 CFR §35.100 et seq.

"Interim controls" means a set of measures designed to reduce temporarily human exposure or likely exposure to lead-based paint hazards including repairs, painting, temporary containment, specialized cleaning, clearance, ongoing lead-based paint maintenance activities, and the establishment and operation of management and resident education programs.

"Lead-based paint" means paint or other surface coatings that contain lead equal to or exceeding 1.0 mg/cm^2 or 0.5% by weight.

"Lead-based paint hazard" means any condition that causes exposure to lead from dust lead hazards, soil lead hazards, or lead-based paint that is deteriorated or present in chewable surfaces, or impact surfaces, and that would result in adverse human health effects.

"Lead-based inspection" means a surface by surface investigation to determine the presence of lead-based paint and provision of a report explaining the results of the investigation.

"Paint stabilization" means repairing any physical defect in the substrate of a painted surface that is causing paint deterioration, removing loose paint and other material from the surface to be treated and applying a new protective coating or paint.

"Paint testing" means the process of determining by a certified lead-based paint inspector or risk assessor the presence or absence of lead-based paint or deteriorated paint surfaces or painted surfaces to be disturbed or replaced.

"Permanent" means an expected design life of at least 20 years.

"Risk assessment" means (1) an onsite investigation to determine the existence, nature, severity and location of lead-based paint hazards and (2) provision of a report by the individual or firm conducting the risk assessment explaining the results of the investigation and options for reducing lead-based paint hazards.

"Soil lead hazard" means bare soil on residential property that contains lead equal to or exceeding levels promulgated by the EPA at 40 CFR 745.65.

"Standard treatments" means a series of hazard reduction measures designed to reduce all lead-based paint hazards in a dwelling unit without the benefit of a risk assessment or other evaluation.

"Substrate" means the material directly beneath the painted surface out of which the components are constructed including wood, drywall, plaster, concrete, brick or metal.

"Zero-bedroom dwelling" means any residential dwelling in which the living areas are not separated from the sleeping area and includes efficiencies, studio apartment, dormitory or single room occupancy housing, military barracks, and rentals of individual rooms in residential dwellings.

HUD's general requirements contain 13 exemptions including residential property for which construction was completed on or after January 1, 1978, zero-bedroom dwelling units, certain housing for the elderly and persons with disabilities, and emergency actions.

The requirements prohibit use of paint containing more than 0.06% dry weight of lead on any interior or exterior surface in federally owned housing or housing receiving federal assistance. They also prohibit several methods of removal of paint that is, or may be, lead-based paint including an unqualified prohibition of open flame burning or torching and qualified prohibitions of other methods unless conditions are met.

The regulations offer some options. Standard treatments may be conducted by a designated party required to conduct interim controls who is willing to presume without evaluation that lead-based paint and lead-based paint hazards or both are present throughout the residential property. If an abatement is required, except for public housing where an inspection is required, a designated party may presume without evaluation that lead-based paint or lead-based hazards or both are present throughout the residential property and conduct abatement on all applicable surfaces including soil. If a risk assessment is required, a designated party may choose first to conduct a lead hazard screen, but must conduct a risk assessment if the results of the screen indicate a need to do so. If paint stabilization or interim controls of deteriorated paint surfaces are required, a designated party may conduct paint testing of all surfaces with non-intact paint and need not stabilize or conduct interim controls for surfaces which the testing indicates an absence of lead-based paint.

Among other things, the regulations contain provisions for notice to occupants of evaluation and hazard reduction activities. They do not relieve anyone from compliance with all applicable federal, state, tribal and local laws. Under some circumstances, an evaluation, hazard reduction or abatement required by the regulations need not be repeated if a prior inspection, risk assessment, interim control program or abatement was performed and met certification and other specified conditions. Records must be kept for three years and records for ongoing maintenance and/or reevaluation activities must be kept until three years after such activities are no longer required.

3. Disposition of Residential Properties Owned by a Federal Agency Other Than HUD

HUD has regulations applicable to sales of residential property owned by a federal agency other then HUD.[17] If construction was before 1960, the federal agency must conduct a risk assessment and a lead-based paint inspection before closing the sale. It must also abate all identified lead-based paint hazards. If abatement is not completed before the closing, it must assure that abatement is carried out by the purchaser before occupancy of the property as target housing. If construction was after 1959 and before 1978, the federal agency is only required to conduct a risk assessment and lead-based paint inspection and make the results available to prospective purchasers.

4. Project-Based Assistance by a Federal Agency Other Than HUD

HUD has regulations applicable to any residential property that receives more than $5,000 annually per project in project-based assistance.[18] Each owner must complete a risk assessment after providing a notice of evaluation or hazard reduction and the EPA lead hazard information pamphlet to occupants. Each owner must then conduct interim controls consistent with the findings of the risk assessment report. If a child under age 6 living in a federally assisted dwelling unit is identified as having an environmental intervention blood lead level, the owner must immediately conduct a risk assessment and then conduct interim controls.

---

[17] 24 CFR §35.200 et seq.

[18] 24 CFR §35.200 et seq.

5. HUD-Owned Single Family Property

HUD has regulations applicable to sales of HUD-owned single-family properties built before 1978 and sold with mortgages insured under a program administered by HUD.[19] The regulations require visual assessment of all painted surfaces to identify deteriorated paint and paint stabilization of all deteriorated paint. Occupancy must not be permitted until all required paint stablization is complete and clearance is achieved. If not completed before the closing, HUD must assure that paint stablization and clearance are carried out by the purchaser before occupancy.

6. Multifamily Mortgage Insurance

HUD has regulations applicable to multifamily residential property for which HUD owns the mortgage or the owner receives mortgage insurance under a program administered by HUD.[20] The regulations exempt an application for insurance in connection with a refinancing transaction where an appraisal is not required under applicable HUD procedures. For property constructed before 1960, before issuance of a firm commitment, the sponsor must conduct a risk assessment and then conduct interim controls to treat lead-based hazards identified in the risk assessment. The sponsor must complete the interim controls before issuance of the firm commitment or they may be made a condition of the commitment with sufficient funds escrowed at initial endorsement of the insured loan. The sponsor must also agree to incorporate ongoing lead-based paint maintenance into regular building operations and maintenance activities. For property constructed after 1959 and before 1978, except for conversions and major rehabilitations, the sponsor must only agree to incorporate ongoing lead-based paint maintenance practices into regular building operations. For a conversion of nonresidential property to residential use or a rehabilitation estimated to cost more than 50% of the estimated replacement cost after rehabilitation, the sponsor must conduct a lead-based paint inspection and, before occupancy, an abatement of all lead-based paint on the property.

---

[19] 24 CFR §35.500 et seq.

[20] 24 CFR §35.600 et seq.

7. Project-Based Assistance

HUD has regulations applicable to certain properties receiving project-based assistance under a HUD program.[21] For multifamily residential properties receiving an average of more than $5,000 per assisted dwelling unit annually in project-based assistance, each owner must complete a risk assessment and then interim controls to treat the lead-based paint hazards identified in the risk assessment. The interim controls must be completed within 90 days in units occupied by families with children under age 6 and related common areas and within 12 months in other dwelling units, common areas and remaining portions of the residential property. Immediately after completion of the risk assessment, the owner must incorporate ongoing lead-based paint maintenance and reevaluation into the regular building operations. If a child under age 6 living in a covered dwelling unit is identified as having an environmental intervention blood level, the owner must comply with risk assessment, verification, hazard reduction, notice and reporting requirements with respect to the dwelling unit where the child lives and the common areas servicing the dwelling unit.

8. HUD-Owned and Mortgage-in-Possession Multifamily Property

HUD has regulations applicable to HUD-owned multifamily residential property or a multifamily residential property for which HUD is mortgagee in possession.[22] HUD must conduct a risk assessment and a lead-based paint inspection before the property is advertised for sale. HUD must then conduct interim controls to treat the lead-based paint hazards identified in the evaluation. The interim controls must be completed within 90 days after completion of the risk assessment in units occupied by families with children under age 6 and within 12 months in other dwelling units. If title conveyance is to occur during those times, HUD must complete the interim controls before conveyance or foreclosure or assure that they are carried out by the purchaser. The purchaser must do so within 180 days, but must do so within 90 days in units occupied by families with children under age 6. If HUD retains ownership of a residential

---

[21] 24 CFR §35.700 et seq.

[22] 24 CFR §35.800 et seq.

property more than 12 months, it must incorporate ongoing lead-based paint maintenance and reevaluation procedures into regular building operation. If a child under age 6 living in a unit is identified as having an environmental intervention blood lead level, HUD must comply with risk assessment, verification, hazard reduction, notice and reporting requirements with respect to the dwelling unit where the child lives and the common areas servicing the unit.

9. Rehabilitation

HUD has regulations applicable to a residential property that receives federal rehabilitation assistance under a program administered by HUD, but not including project-based rental assistance, rehabilitation mortgage insurance or assistance to public housing.[23] All grantees or participating jurisdictions must either perform paint testing or presume that all painted surfaces to be disturbed or replaced are coated with lead-based paint. The extent of compliance depends on the amount of assistance per unit, i.e., (i) up to $5,000 per unit, (ii) more than $5,000 to $25,000 per unit, and (iii) more than $25,000 per unit. Each grantee or participating jurisdiction receiving up to $5,000 per unit must implement safe work practices during rehabilitation work, unless testing shows that paint surfaces to be disturbed are not lead-based paint, and must perform a clearance examination after completion of rehabilitation disturbing lead-based paint surfaces. Each grantee or participating jurisdiction receiving over $5,000 and up to $25,000 per unit must perform a risk assessment, perform interim controls, implement safe work practices, and repair any disturbed paint that is known or presumed to be lead-based paint. Each grantee or participating jurisdiction receiving over $25,000 per unit must perform a risk assessment, abate all lead-based paint hazards identified by the paint testing or risk assessment with some exceptions, implement safe work practices during the rehabilitation work, and repair any disturbed paint that is known or presumed to be lead-based paint. The grantee or participating jurisdiction under the HOME program must require the property to incorporate ongoing lead-based paint maintenance activities in regular building operations. Special requirements apply to insular areas.

---

[23] 24 CFR§35.900 et seq.

10. Acquisition, Leasing, Support Services, or Operation

HUD has regulations applicable to residential properties that receive federal assistance under certain HUD programs for acquisition, leasing, support services and operation, but not including mortgage insurance, sale of federally owned housing, project-based or tenant-based rental assistance, rehabilitation assistance, or assistance to public housing.[24] Each grantee or participating jurisdiction must conduct a visual assessment of all painted surface to identify deteriorated paint; conduct paint stabilization of each deteriorated paint surface and clearance; require the incorporation of ongoing lead-based paint maintenance activities into regular building operations if the dwelling unit has a continuing active financial relationship with a federal housing assistance program; and provide notice to the occupants describing the results of the clearance examination. Except for visual assessment, the grantee or participating jurisdiction must provide notice to residents of each evaluation or hazard reduction and must provide the lead hazard information pamphlet to residents. The grantee or participating jurisdiction must determine whether the cost of evaluation and hazard reduction is to be borne by the owner/developer, the grantee or a combination of them.

11. Public Housing Programs

HUD has regulations applicable to residential property assisted under the U.S. Housing Act of 1937, but not including Section 8 of that Act.[25] After notice to residents and providing the lead hazard information pamphlet, the public housing agency (PHA) must conduct a lead-based paint inspection. If the inspection finds lead-based paint or no inspection has been conducted, the PHA must conduct a risk assessment. Construction contracts, except for emergency work, may not be executed until an inspection and, if required, a risk assessment has been performed. The PHA must then abate lead-based paint and lead-based paint hazards during the course of the physical improvements conducted under the modernization. Where all abatement has not yet occurred, the PHA must conduct interim controls of the lead-based paint

---

[24] 24 CFR §35.1000 et seq.

[25] 24 CFR §35.1100 et seq.

hazards identified in the most recent risk assessment within 12 months after completion of the evaluation, but must do so within 90 days in units occupied by a child under age 6. The PHA must incorporate ongoing lead-based paint maintenance and reevaluation activities into regular building operations unless the building is unoccupied and will be demolished. If a child under age 6 living in a public housing development is identified as having an environmental intervention blood lead level, the PHA must comply with risk assessment, verification, hazard reduction, notice and reporting requirements with respect to the unit where the child lives. If the risk assessment identifies lead-based paint and previous evaluations did not identify them, the PHA must conduct a risk assessment of other units of the building and must conduct interim controls of identified hazards.

12. Tenant-Based Rental Assistance

HUD has regulations applicable to housing occupied by families receiving tenant-based rental assistance including the Section 8 certificate program, the Section 8 voucher program, the HOME program, the Shelter Plus Care Program, the Housing Opportunities for Persons with AIDS (HOPWA) program, and the Indian Housing Block Grant Program. However, applicability is limited to dwelling units occupied or to be occupied by families or households with one or more children under age 6 and associated common areas and exterior painted surfaces. During initial and periodic inspections, the designated party under the program must arrange a visual inspection by a trained inspector to identify deteriorated paint. After providing notice and the lead hazard information pamphlet to residents, the owner must stabilize each deteriorated paint surface before commencement of assisted occupancy. If assisted occupancy previously commenced, paint stabilization must be completed within 30 days of notification of the owner of the results of the visual assessment. For certain programs, the owner must incorporate ongoing lead-based paint maintenance activities into regular building operations. If a child under age 6 living in an assisted dwelling unit is identified as having an environmental intervention blood lead level, the designated party must comply with risk assessment, verification, hazard reduction, notice and reporting requirements with respect to the dwelling unit where the child lived and the common areas servicing the dwelling unit.

THE LEAD-BASED PAINT POISONING PREVENTION ACT 42 U.S.C. §4801, AND THE RESIDENTIAL LEAD-BASED PAINT REDUCTION ACT, 42 U.S.C. §4851

13. Methods and Standards for Evaluation and Hazard Reduction

HUD's regulations provide methods and standards for evaluation and hazard reduction activities.[26]

The regulations refer for further guidance to the HUD Guidelines for the Evaluation and Control of Lead-Based Paint Hazards in Housing; the EPA Guidance on Residential Lead-Based Paint, Lead-Contaminated Soil; and guidance methods or protocols issued by States and Indian Tribes authorized by EPA to administer and enforce lead-based paint programs. HUD's Guidelines are available on its website.

Paint chip, dust and soil samples must be collected and analyzed in accordance with standards established by a State or Indian Tribe or under a program authorized by the EPA.

Lead-based paint inspections, risk assessments, lead hazard screens and reevaluations must be performed in accordance with methods and standards established either by a state or tribal program authorized by the EPA or established by the EPA. Paint testing must be performed by a certified lead-based paint inspector or risk assessor. Reevaluation must be performed by a certified risk assessor.

Risk assessors must use standards at least as protective as those promulgated by the EPA under TSCA. HUD's dust lead standard for (1) floors is 40 mg/m^2 for risk assessment, reevaluation and clearance; (2) interior window sills is 250 mg/m^2; and (3) window troughs is not applicable for risk assessment and reevaluation and is 400 mg/m^2 for clearance. HUD's soil lead hazard standard for bare soil in play areas of children under age 6 is 400 ug/g (400 ppm) and for bare soil in the rest of the yard is 1,200 ug/g (1,200 ppm).

Abatement must be performed in accordance with methods and standards established either by a State or Indian Tribe under a program authorized by the EPA or established by the EPA. If encapsulation or enclosure is used as a method of abatement, ongoing lead-based paint maintenance activities must be performed. Abatement of an intact, factory-applied prime coating on metal surfaces is not required unless the surface is a friction surface.

[26] 24 CFR §35.1300 et seq.

Interim controls of lead-based paint hazards identified in a risk assessment include paint stabilization of deteriorated paint, treatment of friction and impact surfaces, dust control, and lead-contaminated soil control and must be conducted in accordance with 24 CFR §35.1330. Interim controls, may be performed in combination with, or be replaced by, abatement methods. Only those interim control methods identified as acceptable in a current risk assessment report may be used, except that a risk assessment is not required if only paint stabilization is required by the regulations. Occupants must be protected during the work and clearance testing must be performed at the conclusion. The person performing the interim controls must be trained in the OSHA construction industry hazard communication standard and either be supervised by a certified lead-based paint abatement supervisor or have successfully completed the EPA's abatement supervisor or worker course or another course approved by the EPA or a course approved by HUD.

Standard treatments must be conducted in accordance with 24 CFR §35.1335. Standard treatments include paint stabilization; covering of smooth and cleanable horizontal surfaces; correcting dust-generating conditions; treating bare residential soil; using safe work practices; performance of a clearance examination; and performance by individuals who meet training and/or supervision requirements.

Clearance examinations must be performed in accordance with 40 CFR §35.1340. They must be performed by a certified risk assessor, a certified lead-based paint inspector, or other qualified personnel who have met training or licensing requirements. The required activities including a visual assessment, dust sampling, submission of samples for analysis for lead in dust; interpretation of the sampling results, and preparation of a report.

Occupant protection and worksite preparation must comply with 24 CFR §35.1345. With some exceptions, occupants must not be permitted to enter the worksite and must be temporarily relocated before and during hazard reduction activities to a suitable, decent, safe and similar accessible dwelling unit that does not have lead-based paint hazards. The dwelling unit and the worksite must be protected against unauthorized entry and occupants belongings protected from contamination. Worksite preparation includes steps to prevent release of leaded dust, contamination of paint chips and other debris within the worksite, and posting of warning signs.

Safe work practices must comply with 24 CFR §35.1350. Prohibited methods of paint removal must not be used. Occupant protection and worksite preparation steps must be implemented and performed or supervised by trained personnel. After hazard reduction activities are completed, the worksite must be cleaned using methods, products and devices that are successful in cleaning up dust-lead hazards such as HEPA vacuums and lead-specific detergents or equivalents. Safe work practices are not required when maintenance or hazard reduction activities do not disturb painted surfaces that total more than de minimis levels: (1) 20 square feet on exterior surfaces, (2) 2 square feet in any one interior room or space, or (3) 10% of the total surface area on an interior or exterior type of component with a small surface area such as window sills, baseboards and trim.

Ongoing lead-based paint maintenance and reevaluation activities must be conducted in accordance with 24 CFR §35.1355. They need not be conducted if a lead-based paint inspection indicates that no lead-based paint is present in the dwelling units, common areas or exterior surfaces or a clearance report indicates that all lead-based paint has been removed. A visual assessment for deteriorated paint, bare soil and failure of any hazard reduction measures shall be performed at unit turnover and every 12 months. Deteriorated paint must be stabilized and bare soil treated with standard treatments using safe work practices and other required methods. Reevaluations must be conducted, with some exceptions, at two year intervals and in certain other events by a certified risk assessor.

F. Other HUD Regulations

Other HUD programs have regulations on lead-based paint that generally refer to several of the requirements of 24 CFR Part 35 summarized earlier in this Chapter. Examples of these cross-referral regulations are as follows:

Federal Housing Administration. 24 CFR §200.77 and .800

Nehemiah Housing Program. 24 CFR §280.25(e)

HUD Single Family Property. 24 CFR §291.100(g) and .430

Rental Rehabilitation Grant Program. 24 CFR §511.15

Community Development Block Grants. 24 CFR §570.461 and .608

HOPE 3 Home Ownership Program. 24 CFR §572.420(h)

Loan Guaranty Recovery Fund. 24 CFR §573.9(c)

Housing for Persons with AIDS. 24 CFR §574.635

Emergency Shelter Grants Program. 24 CFR §576.57(c)

Supportive Housing Program. 24 CFR §583.330(d)

Youthbuild Program. 24 CFR §585.502(h)

Urban Empowerment Zones.24 CFR §598.408

Congregate Housing Services Program. 24 CFR §761.40(c)

Section 8 Housing Assistance Payments Program. 24 CFR §881.207(e)

Section 8 Moderate Rehabilitation Programs. 24 CFR §882.404(d)

Section 8 Housing Assistance Program. 24 CFR §883.310(b)(5)

Section 8 Housing Assistance Program. 24 CFR §886.113(i)

Supportive Housing for the Elderly. 24 CFR §891.335

Public Housing Home Ownership Programs. 24 CFR §906.7

Indian Home Program. 24 CFR §954.4(g)

PHA Owned or Leased Projects. 24 CFR §965.215 (Insurance)
and 24 CFR §965.701

Public Housing Modernization. 24 CFR §968.110(k)

Native American Housing Activities. 24 CFR §1000.40

G. Litigation

In recent years, several lawsuits under the Poisoning Prevention Act and the Residential Hazard Reduction Act have resulted in published court decisions.

Three trade associations challenged the "disclosure rule" adopted by HUD and the EPA which requires disclosure to a purchaser or lessee of target housing of the presence of any known lead-based paint and/or lead-based paint hazards and defines lead-based paint hazards to include all residential lead-containing dusts and soils regardless of source. The court held that the definition was rational, given the impossibility of determining whether lead contamination in soil and dust derived from paint or elsewhere. National Multi-Housing Council v. U.S.E.P.A., 292 F.3d 232 (D.C. Cir. 2002).

The Court of Federal Claims lacks jurisdiction over claims against the United States under the Residential Hazard Reduction Act and TSCA. Ayres v. US., 67 Fed. Cl. 776 (2005).

The waiver of sovereign immunity in TSCA does not extend to the Residential Hazard Reduction Act so as to allow a private action against the Department of Veterans Affairs for treble damages under that Act. Cudjoe v. Dept of Veterans Affairs, 426 F. 3d 241 (3d Cir. 2005).

Minor children of a lessee do not have standing to sue lessors alleging failure to disclose information required by the Residential Hazard Reduction Act. Mason ex rel. Heiser v. Morrisette, 403 F.3d 28 (1st Cir. 2005).

A buyer has no cause of action against her own broker or agent for alleged failure to ensure compliance with the Residential Hazard Reduction Act. Flowers v. ERA Unique Real Estate, Inc., 170 F. Supp. 2d 840 (N.D. Ill. 2001). See also Keegan v. Downing Agency, Inc., 2003 WL 21210326 (D. Me., May 22, 2003) and Griffin v. Bruner, 793 N.E.2d 974 (Ill. App. 2003).

A lessee lacks standing to sue a real estate agent who administered the sale of a home for HUD that was subsequently leased to the plaintiff lessee. Wallace v. U.S., 335 F. Supp. 2d 252 (D.R.I. 2004).

The U.S. Circuit Court of Appeals for the Third Circuit and some other courts found an implied private right of action under Poisoning Prevention Act and the federal housing laws. Davis v. Philadelphia Housing Authority, 121 F.3d 92 (3d Cir. 1997); Aristel v. Housing Authority, 54 F. Supp.2d 1289 (M.D. Fla. 1999); Sipes v. Russell, 89 F. Supp.2d 1199 (D. Kans.

2000). However, other courts have held to the contrary. Santiago v. Hernandez and the City of New York, 53 F. Supp.2d 264 (E.D.N.Y. 1999).

The civil rights provisions of the U.S. Housing Act, §1983 cannot be combined with the Poisoning Prevention Act and the Paint Hazard Reduction Act to create private rights of action to sue municipalities and their agencies for alleged injuries from lead-based paint. Johnson v. City of Detroit, 319 F. Supp. 2d 756 (E.D. Mich. 2004); aff'd 446 F.3d 614 (2006); LB, III v. Housing Authority of Louisville, 345 F. Supp. 2d 725 (W.D. Ky, 2004); Mair v. City of Albany, 303 F. Supp. 2d 237 (N.D.N.Y. 2004).

Sellers of a residence with no actual knowledge of lead paint hazards who complied with the EPA's lead-based paint disclosure regulations were not required by the Residential Hazard Reduction Act or common law to inspect for lead-based paint before a sale. Nunez v. J.L. Sims Co., 2003 WL 2143328 (Ohio Ct. App. June 27, 2003). However, a partial summary judgment was granted against sellers who failed to deliver a lead-based paint report before execution of the contract for sale, even though they delivered the report at the closing of the contract. Smith v. Coldwell Banker, 122 F. Supp. 2d 267 D. Conn. 2000).

XVII. THE TOXIC SUBSTANCES CONTROL ACT

A. General Provisions of TSCA

The Toxic Substances Control Act (TSCA) applies to persons who manufacture, import, process, or distribute chemical substances and mixtures and also to persons who dispose of some chemical substances.[1] TSCA is administered by the EPA.

TSCA requires a 90-day pre-manufacturing notice (PMN) to the EPA before manufacturing, importing or processing of a new chemical substance. A new chemical substance is any not included in an inventory over 80,000 chemical substances maintained by the EPA.[2] TSCA also authorizes the EPA to require a significant new use notice (SNUN) before manufacturing, importing or processing any chemical substance for a use determined by the EPA to be a significant new use. The EPA's significant new use rules are called SNURs.[3]

TSCA authorizes the EPA to require testing of chemical substances or mixtures. The EPA's test guidelines are contained in 40 C.F.R. Parts 795 through 799. TSCA authorizes the EPA to impose restrictions on a chemical substance or mixture if it finds there is reasonable basis to conclude that its manufacture, processing, distribution, use or disposal will result in an unreasonable risk of injury to health or the environment. The restrictions may include prohibition of the substance or mixture, limitation on concentrations in manufacture or use, warning labels and instructions, recordkeeping, limitations on use, prohibition or regulation of disposal, and recall or repurchase.

TSCA requires detailed recordkeeping and reporting. For example, records of "significant adverse reactions" alleged to have been caused by a chemical substance or mixture must be made. Allegations made by or on behalf of employees must be recorded and retained for 30 years and those by or on behalf of other persons for 5 years. A "significant adverse reaction" means a reaction that may indicate a substantial impairment of normal activities, or long-lasting or irreversible damage to health or the environment. The EPA has provided rules clarifying

---

[1] 15 U.S.C. § 2601 et seq.

[2] 40 C.F.R. Parts 710 and 720.

[3] 40 C.F.R. Part 721.

these recordkeeping and reporting obligations.[4] New health and safety studies must also be reported for a list of substances which includes lead naphthenate.[5]

Section 8(e) of TSCA requires that any person who manufactures, imports, processes or distributes a chemical substance or mixture and who obtains information which reasonably supports that it represents a substantial risk of injury to health or the environment shall immediately inform the EPA unless such person has actual knowledge that the EPA has been adequately informed of such information. The EPA allows 30 days to submit such information.

B. Advance Notice of Proposed Rulemaking for a Wide Review of Lead in the Environment under TSCA

In 1991, the EPA published an advance notice of proposed rulemaking (ANPR) stating that it was exploring the desirability and feasibility of various approaches under TSCA to significantly reduce any unreasonable risks to human health and the environment from exposures to lead and lead compounds.

The EPA acknowledged that the average blood lead level in U.S. children had declined from about 17 ug/dL to between approximately 4 and 6 ug/dL over the past decade. However, the EPA described a spectrum of human health effects and environmental effects that might justify further regulation including (1) a phaseout of certain uses of lead such as brass and bronze plumbing fittings and fixtures (e.g., faucets); (2) a general reduction of lead use; (3) regulations encouraging recycling of products in addition to lead-acid batteries, perhaps by restricting production of virgin lead; and (4) product stewardship programs from mining to disposal. Since 1991, the EPA has not adopted regulations directly resulting from the ANPR. However, some of the regulatory programs discussed in the ANPR have become part of other regulatory programs of the EPA and of state and local programs.

---

[4] 40 C.F.R. Part 717.

[5] 56 FR 22096 (May 13, 1991).

C. Proposed Ban on Lead and Zinc Fishing Sinkers.

In 1992, four environmental groups petitioned the EPA to require that sales of lead fishing sinkers be labelled to warn that such products are toxic to wildlife. Not satisfied with the remedy proposed by the environmental groups, the EPA published in 1994 a proposed rule to prohibit the manufacturing, processing and distribution in commerce in the United States of fishing sinkers containing lead or zinc, including brass, having any dimension equal to or less than one inch. Exporters of such fishing sinkers would be required to notify the EPA which will notify environmental officials in the foreign country of destination.[6] Since 1994, the EPA has not adopted the proposed rule and its website states that the EPA intends to withdraw the proposal.

D. Proposed Significant New Use Regulations

In 1994, the EPA issued an ANPR under TSCA stating that it was exploring the possibility of issuing SNURs for certain uses of elemental lead and lead compounds. The SNURs would require persons to notify the EPA at least 90 days before commencing manufacture, import or processing of lead for uses identified by the EPA as significant new uses.[7] Since 1994, the EPA has not issued SNURs in the manner contemplated by the ANPR, but has accomplished some of the regulatory objectives of the ANPR in other regulations which it adopted under TSCA and other laws.

E. TSCA Programs for Lead-Based Paint

Title X of the Residential Lead-Based Paint Hazard Reduction Act of 1992 added a new Subchapter IV to TSCA assigning numerous tasks relating to lead-based paint to the EPA and other federal government agencies.[8] Lead-based paint was defined as any paint or other surface coating containing lead in excess of 1.0 mg/cm^2 or 0.5% by weight or such lower levels in target housing as may be established by HUD or for other coatings by the EPA.

---

[6] 59 Fed. Reg. 11122 (March 9, 1994).

[7] 59 Fed. Reg. 49484 (Sept. 28, 1994).

[8] 15 U.S.C. §2681 et seq.

The new Subchapter IV required that the EPA, in consultation with OSHA, HUD, DHHS and NIOSH, promulgate regulations governing lead-based paint activities to ensure that individuals engaged in such activities are properly trained; training programs are accredited; and contractors engaged in such activities are certified. The regulations must require that risk assessment, inspection, and abatement activities performed in "target housing" (i.e., all housing constructed before 1978, with certain exceptions) must be performed by certified contractors.[9]

Important provisions of the Act are summarized as follows:

- Accreditation of Training Programs. The EPA was required to adopt regulations containing specific requirements for accreditation of lead-based paint activities training programs for workers, supervisors, inspectors, planners and other involved individuals. The regulations were required to contain minimum requirements for accreditation of training providers, training curriculum, training hours, hands-on training, training competency and proficiency, and training program quality control.[10]
- Accreditation and Certification Fees. With certain exceptions, the EPA (or the state in an authorized state program) was required to impose fees on persons operating accredited training programs and certified lead-based paint activities contractors to cover the costs of administering and enforcing the standards and regulations applicable to them.[11]
- Renovation and Remodeling. The EPA was required to issue guidelines for renovation and remodeling of target housing, public buildings constructed before 1978, and commercial buildings as to those activities which may create risk of exposure to dangerous levels of lead, and to distribute such guidelines to persons engaged in renovation and

---

[9] 15 U.S.C. § 2682.

[10] 15 U.S.C. § 2682(a)(1) and (2).

[11] 15 U.S.C. § 2682(a)(3).

remodeling through hardware and paint stores, employee organizations, trade groups, state and local agencies, and other means. The EPA was also required to adopt regulations applicable to renovation or remodeling activities in target housing, public buildings constructed before 1978, and commercial buildings that create lead-based paint hazards, except insofar as the EPA might find and explain its reasons for exceptions.[12]

- Identification of Dangerous Levels of Lead. The EPA was required to adopt regulations identifying lead-based paint hazards, lead-contaminated dust, and lead-contaminated soil.[13]
- Authorized State Programs. States were allowed to apply to the EPA to administer and enforce the standards, regulations and other requirements applicable to training and certification and a lead hazard information pamphlet. The EPA may approve a state application only if it finds, after notice and opportunity for public hearing, that the state program is at least as protective of human health and the environment as the federal program and provides adequate enforcement. The EPA was required to promulgate a model state program. The EPA was authorized to make grants to the states to develop and carry out their programs and to withdraw authorization for noncompliance. If a state did not have an authorized state program within two years after promulgation of the EPA's regulations, the EPA was required to establish a federal program in the State.[14]
- Lead Abatement and Measurement. The EPA and other federal agencies were directed to conduct a comprehensive program to promote safe,

---

[12] 15 U.S.C. § 2682(c).

[13] 15 U.S.C. § 2683.

[14] 15 U.S.C. § 2684.

effective and affordable monitoring, detection and abatement of lead-based paint and other lead exposure hazards. The EPA was required to establish protocols, criteria, and minimum performance standards for analysis of lead in paint films, soil and dust. The EPA was also required to establish a program to certify laboratories as qualified to test substances for lead content and to publish an annual list of certified or accredited laboratories. The CDC and NIEHS were directed to conduct joint studies of the sources of lead exposure in children with elevated blood levels. The EPA, in consultation with NIOSH, was directed to conduct a comprehensive study of means to reduce hazardous occupational lead abatement exposures. The EPA was directed to coordinate with the education and awareness program on lead poisoning of the President's Council on Environmental Quality. The EPA, in consultation with the CPSC, was directed to develop and distribute to consumers information related to hazards of renovation and remodeling where lead-based paint may be present. The EPA and other federal Agencies were directed to establish a National Clearinghouse on Childhood Lead Poisoning and a "hotline" to provide answers to public questions. The President was directed to cause the EPA, the Department of Commerce or another agency to establish criteria, testing protocols, and performance characteristics for new hazard evaluation and reduction products.[15]

- Lead Hazard Information Pamphlet. The EPA, in consultation with HUD and the DHHS, was required to publish a lead hazard information pamphlet. The contents of the pamphlet were prescribed in some detail. The EPA was required to adopt regulations requiring each person who performs for compensation a renovation of target housing to provide the

[15] 15 U.S.C. § 2685.

pamphlet to the owner and occupant of such housing prior to commencing the renovation.[16]

- Control of Lead-Based Paint Hazards at Federal Facilities. Federal government departments, agencies and instrumentalities were required to comply with federal, state, interstate and local requirements respecting lead-based paint, lead-based paint activities and lead-based paint hazards and are subject to orders, fines and penalties. A waiver of the immunity of the United States was provided. However, the Act provided that no agent, employee of officer of the United States shall be personally liable.[17]

The Act also required HUD and the EPA to issue regulations under TSCA requiring that every seller and lessor of "target housing" meet several requirements for the benefit of prospective buyers or tenants. "Target housing" was defined as any housing constructed prior to 1978 except certain housing for the elderly and disabled and any 0-bedroom dwelling such as the typical hotel or motel room. The Act said that regulations must require sellers and lessors to furnish a lead hazard information pamphlet prepared by the EPA, disclose the presence of any known lead-based paint or hazards in the housing, and permit a 10-day period to conduct a risk assessment or inspection. The regulations must require that contracts for purchase or sale of any interest in target housing contain a warning statement and a statement signed by the purchaser that evidences compliance with the requirements of TSCA.

F. Lead-Based Paint Hazards.

1. Early Guidance (1994)

On July 14, 1994, the EPA released a memorandum (with enclosures) titled "Guidance on Residential Lead-Based Paint, Lead-Contaminated Dust, and Lead Contaminated Soil" pursuant to Section 403 of Title IV of TSCA. The guidance information was intended to

---

[16] 15 U.S.C. § 2686.

[17] 15 U.S.C. § 2688.

serve until the EPA could propose and adopt regulations on the subject. The guidance was later published in the Federal Register under the title "Guidance on Identification of Lead-Based Paint Hazards."[18] In the Guidance, the EPA recommended that the following clearance levels for lead-contaminated dust be met after abatement or interim control activities have been performed.

Location	Lead Loading
Uncarpeted floors	100 ug/ft^2 (0.93 mg/m$_2$)
Interior window sills	500 ug/ft^2 (4.65 mg/m$_2$)
Window wells	800 ug/ft^2 (7.45 mg/m$_2$)

The EPA recommended that lead-contaminated soil be evaluated based on lead concentration levels. At levels below 400 ppm, the Guidance did not recommend action. When lead concentrations exceed 400 ppm in base soil, the EPA recommended that further evaluation be undertaken as well as physical exposure reduction activities commensurate with risk. Within the range of 400 ppm and 4,000 ppm, the EPA recommended risk reduction activity, primarily based on exposure of children. However, where bare-soil lead levels were 2,000 ppm or more, the EPA recommended that internal controls be implemented even if the area was not frequented by children. At 5,000 ppm or more, the EPA recommended soil abatement including removal or replacement of soil or the use of permanent covers such as paving.

2. The Lead-Based Paint Hazards Regulation (2001)

On January 5, 2001, the EPA published a final regulation establishing standards determining lead-based paint hazards for paint, dust and soil.[19]

A "paint-lead hazard" is any of the following:

(a) Any lead-based paint on a friction surface that is subject to abrasion and where the lead dust levels on the nearest horizontal surface underneath the friction surface (e.g., the window sill, or floor) are equal to or greater than the dust-lead hazard levels identified as a dust-lead hazard below.

[18] 60 Fed. Reg. 47248 (Sept. 11, 1995).

[19] 40 CFR § 745.65.

(b) Any damaged or otherwise deteriorated lead-based paint on an impact surface that is caused by impact from a related building component such as a doorknob that knocks into a wall or a door that knocks against its doorframe.

(c) Any chewable lead-based painted surface on which there is evidence of teeth marks.

(d) Any other deteriorated lead-based paint in any residential building or child-occupied facility or on the exterior of any residential building or child-occupied facility.

A "dust lead hazard" is surface dust in a residential dwelling or child-occupied facility that contains a mass-per-area concentration of lead equal to or exceeding 40 ug/ft^2 on floors or 250 ug/ft^2 on interior windows based on wipe samples.

A "soil lead hazard" is bare soil on a residential real property or on the property of a child-occupied facility that contains total lead equal to or exceeding 400 parts per million in a play area or average of 1,200 parts per million of bare soil in the rest of the yard based on soil samples.

The EPA's adoption of a regulation defining lead-based paint hazards as including dust-lead and soil-lead hazards regardless of whether attributable to lead-based paint was upheld in National Multi Housing Council v. U.S. Environmental Protection Agency, 292 F.3d 232 (D.C. Cir. 2002) on grounds that under current technology it is impossible to ascertain whether lead contamination in soil and dust derives from lead-based paint or some other source.

G. The Residential Property Renovation Regulations (1998)

In 1998, the EPA adopted regulations under Section 402(c)(1) of TSCA to ensure that owners and occupants of "target housing" understand the potential hazards of lead-based paint exposure before certain renovations are started.[20]

The renovation regulations apply to all renovations of target housing performed for compensation. Their purpose is to require renovators to provide a lead hazard information

[20] 40 CFR § 745.80.

pamphlet to the owner and occupant of such housing before commencing renovation. The regulations do not apply to (1) minor repair and maintenance activities that disrupt two square feet or less of painted surface per component, (2) emergency renovation operations, or (3) renovations in target housing in which a properly certified inspector has made a written determination that the components affected by the renovation are free of paint or other surface coatings that contain lead equal to or in excess of 1.0 mg/cm^2 or 0.5% by weight and the renovator has obtained a copy of the determination.

The renovation regulations contain several definitions including a broad definition of "renovation" which means modification of any existing structure, or portion thereof, that results in disturbance of painted surfaces unless that activity is performed as part of an abatement as defined by the EPA's lead-based paint activity regulations described later in this Chapter. The definition warns that renovation includes, without limitation, removal or modification of painted surfaces or painted components such as modification of painted doors; surface preparation activities such as sanding or scraping; removal of large structures such as walls and ceilings; large surface replastering or major replumbing; and window replacement.

The renovation regulations set procedures for delivery of the EPA's information pamphlet, currently called "Protect Your Family from Lead in Your Home" no more than 90 days before beginning renovation activities in any residential unit of target housing. The renovator must provide the pamphlet to the unit owner and either obtain from the owner a written acknowledgement of its receipt or obtain a certificate of mailing at least 7 days before the renovation. If the owner does not occupy the dwelling unit, the renovator must either (1) obtain a written acknowledgement of receipt of the pamphlet from an adult occupant or certify in writing that a pamphlet was delivered and the renovator was unsuccessful in obtaining a written acknowledgement from an adult occupant or (2) obtain a certificate of mailing at least 7 days before the renovation. The regulations prescribe similar methods for common areas.

The renovation regulations prescribe the contents of written acknowledgments and certifications of attempted delivery and also provide sample acknowledgment and certification language.

Sample acknowledgment language is as follows:

I have received a copy of the pamphlet, *Protect Your Family From Lead in Your Home*, informing me of the potential risk of lead hazard exposure from renovation activity to be performed in my dwelling unit. I received this pamphlet before the work began.

____

____

Printed Name and Signature

____

Date

____

Unit Address

Sample certification language is as follows:

(i) Unavailable for signature.

I certify that I have made a good faith effort to deliver the pamphlet, *Protect Your Family From Lead in Your Home*, to the unit listed below and that the occupant was unavailable to sign the acknowledgment. I further certify that I have left a copy of the pamphlet at the unit by sliding it under the door.

____

____

Printed Name and Signature

____

Date

____

Unit Address

(ii) Refusal to sign.

I certify that I have made a good faith effort to deliver the pamphlet, *Protect Your Family From Lead in Your Home*, to the unit listed below at the dates and times indicated, and that the occupant refused to sign the acknowledgment. I further certify that I have left a copy of the pamphlet at the unit with the occupant.

____

____

Printed Name and Signature

____

Date

____

Unit Address

Attempted delivery dates and times:

Copies of the pamphlet can be obtained from the EPA's website and the National Lead Information Center Clearinghouse (1-800-424-LEAD). A Spanish-language version of the pamphlet is also available.

In 2006, the EPA published a proposed rule that would make important amendments to the renovation regulations. It is described later in this Chapter.

H. The Disclosure Regulations (1994)

In 1994, HUD and the EPA jointly proposed a disclosure rule to implement amendments to Section 1018 of the Residential Lead-Based Paint Hazard Reduction Act of 1992. The Regulations, adopted June 1, 1998, appear at 40 CFR § 745.100 to § 745.119 under the title "Disclosure of Known Lead-Based Paint and/or Lead-Based Paint Hazards Upon Sale or Lease of Residential Property." The regulations do not apply to foreclosure sales, leases of target housing found to be lead free by a certified inspector, short term leases of 100 days or less where no renewal or extension can occur, and lease renewals if the tenant has already received all information known to the landlord. They also exclude "0-bedroom dwellings" in which there is no separation of the living and bedroom area (such as single room efficiency units and most hotel and motel rooms) and certain housing for the elderly.

Sellers and lessors (and their agents, if any) must provide to purchasers and lessees of target housing all information known to them on the presence of lead-based paint and lead-based paint hazards in the housing and also an EPA pamphlet describing such hazards (i.e., "Protect Your Family from Lead in Your Home") before the purchaser or lessee is obligated under any contract to purchase or lease the housing.

In addition, sellers of target housing must grant 10 calendar days to any prospective purchaser who submits a written purchase offer to conduct an inspection or risk assessment for lead-based paint hazards before being obligated under any contract to purchase target housing. However, a purchaser may waive the opportunity to conduct the risk assessment or inspection by so indicating in writing.

Standard warning, disclosure and acknowledgement forms must be signed, dated and attached to all sale and lease contracts. Records evidencing fulfillment of the disclosure requirements must be kept for 3 years by sellers, lessors and their agents.
If a contract is written in a language other than English, it must include a translation of the lead warning statement. The Government Printing Office publishes a translation in Spanish.

The lead warning statement for a contract to sell reads as follows:

> Every purchaser of any interest in residential real property on which a residential dwelling was built prior to 1978 is notified that such property may present exposure to lead from lead-based paint that may place young children at risk of developing lead poisoning. Lead poisoning in your children may produce permanent neurological damage, including learning disabilities, reduced intelligence quotient, behavioral problems and impaired memory. Lead poisoning also poses a particular risk to pregnant women. The seller of any interest in residential real property is required to provide the buyer with any information on lead-based paint hazards. A risk assessment or inspection for possible lead-based paint hazards is recommended prior to purchase.

The lead warning statement for lessor reads as follows:

> Housing built before 1978 may contain lead-based paint. Lead from paint, paint chips, and dust can pose health hazards if not managed properly. Lead exposure is especially harmful to young children and pregnant women. Before renting pre-1978 housing, lessors must disclose the presence of lead-based paint and/or lead-based paint hazards in the dwelling. Lessees must also receive a federally approved pamphlet on lead poisoning prevention.

The regulations make agents acting for sellers or lessors responsible to assure compliance by sellers and lessors for whom they are acting as well as for recordkeeping requirements. The regulations require sellers and lessors to disclose to agents the presence of any known lead-based paint or lead-based paint hazards and any additional information known to them about lead-based paint and lead-based paint hazards.

The definition of a lead-based paint hazard is very broad. It means any condition that causes exposure to lead from lead-contaminated dust, lead-contaminated soil, lead-contaminated paint that is deteriorated or present in accessible surfaces, friction surfaces or impact surfaces that would result in adverse human health effects as established by the appropriate Federal Agency. Thus, lead-based paint on "chewable" surfaces (such as window sills) and on surfaces often kicked (such as stair risers) may be regarded as a hazard.

Consistent with the Act, the EPA's regulations do not require that:

1. Sellers or lessors inspect for lead-based paint. Thus, most sellers and lessors have no information to disclose because lead-based paint has not been used in residential housing for several decades.

2. Purchasers or lessees exercise their right to an inspection or risk assessment.

3. Sellers or lessors remove or otherwise abate lead-based paint or lead-based paint hazards, if identified by an inspection or risk assessment, or agree to any requested price adjustment.

As described in other Chapters, removal or other abatement of lead-based paint is not easy. Costs may be several thousand dollars per room and waiting times may be several weeks. Thus, many buyers waive the right to an inspection or risk assessment. They know or learn that good maintenance, housekeeping and supervision of children can prevent lead-based paint related health problems. The EPA itself has stated that housing containing lead-based paint, if properly managed and maintained, poses little risk. (See EPA's Qs and As, EPA-747-f-94, October 1994.)

I. Lead-Based Paint Activities Regulations (1994)

In 1994, the EPA published proposed regulations under Section 402(a)(1) of TSCA to govern lead-based paint activities, including accreditation of training programs; certification of individuals and firms engaged in lead-based paint activities; and standards for conducting lead-based paint activities. In 1996, the final regulations were adopted.[21]

---

[21] 59 Fed. Reg. 45872 (Sept. 2, 1994); 61 Fed. Reg. 45825 (Aug. 29, 1996); 40 CFR § 745.239.

In the same 1994 announcement, the EPA published proposed regulations for state programs on lead-based paint activities and also for programs of Indian Tribes. The final regulations were adopted in 1996.[22] They are described later in this Chapter.

The activities regulations do not require any owner or occupant to undertake lead-based paint activities. Rather, they establish requirements and procedures applicable to lead-based paint activities, including individuals and firms engaged in lead-based activities and work practice standards for performing such activities.

Lead-based paint activities are defined in the regulations as inspection, risk assessment and abatement. Such activities must be conducted only according to the procedures and work practice standards prescribed in the activities regulations. No individual or firm may offer to perform any such activity unless certified to perform the activity.

An inspection means a surface-by-surface investigation to determine the presence of lead-based paint and provision of a report explaining the results of the investigation.

A risk assessment means an on-site investigation to determine the existence, nature, severity, and location of lead-based paint hazards and the provision of a report by the individual or the firm conducting the risk assessment explaining the results of the investigation and options for reducing lead-based paint hazards.

Abatement is defined to mean any measure or set of measures designed to permanently eliminate lead-based paint hazards. Abatement does not include renovation, remodeling, landscaping or other activities not designed to permanently eliminate lead-based paint, but instead to repair, restore, or remodel a structure or dwelling, even though these activities may incidentally result in a reduction or elimination of lead-based paint hazard. Abatement also does not include interim controls, operations and maintenance activities, or other measures and activities designed to temporarily, but not permanently, reduce lead-based paint hazards. The EPA has said that abatement is not the only option for controlling lead-based paint hazards and is not always the best or most appropriate response to them. The EPA has described effective control measures as including encapsulation, enclosure and permanent replacement.

---

[22] 59 Fed. Reg. 45872 (Sept. 2, 1994); 61 Fed. Reg. 45825 (Aug. 29, 1996); 40 CFR § 745.320 to 745.339.

The regulations do not apply to persons who perform lead-based paint activities at residential dwellings which they own, unless the residential dwelling is occupied by a person or persons other than the owner's immediate family while the activities are being conducted or a child residing in the building has an elevated blood lead level. The EPA interprets Section 402(a) of TSCA as requiring that a homeowner who hires a contractor to do abatement in his home must hire a certified contractor.

The regulations on accreditation of training programs prescribe: (1) the application process to obtain accreditation; (2) the requirements for the accreditation of training programs; (3) minimum training curriculum requirements; (4) requirements for the accreditation of refresher training programs; (5) reaccreditation of training programs; (6) suspension, renovation and modification of accreditated training programs; and (8) training programs recordkeeping requirements.[23] Training providers may not provide, or claim to provide, EPA accredited lead-based paint activities courses without applying for and receiving accreditation from the EPA.[24]

The regulations on certification of individuals and firms engaged in lead-based paint activities state that it is a violation of TSCA for an individual to conduct any lead-based paint activities unless certified by the EPA pursuant to the regulations.[25] The regulations provide for certification of individuals in specified categories: (1) inspector, risk assessor or supervisor, and (2) abatement worker and project designer. For some individuals, certification may be based on prior training. Recertification is required every three years or five years, based on completion of refresher courses. The regulations provide for certification of firms. They also provide for suspension, revocation and modification of certifications of individuals and firms engaged in lead-based paint activities.[26]

---

[23] 40 CFR § 745.225.

[24] 40 CFR § 745.225(a)(3).

[25] 40 CFR § § 745.226(a)(5).

[26] 40 CFR § § 745.226(g) and (h).

The regulations on standards for conducting lead-based paint activities require that all such activities be performed pursuant to the work practice standards.[27] The work practice standards for lead-based paint activities in target housing and child occupied facilities are quite detailed and include requirements for (1) inspections; (2) lead hazard screens; (3) risk assessments; (4) abatement; (5) collection and laboratory analysis of samples; (6) composite dust sampling; (7) determinations of the presence of lead-based paint, a paint lead hazard, a dust-lead hazard and a soil-lead hazard; and (8) recordkeeping. The regulations state that documented methodologies appropriate for compliance are found in the HUD Guidelines for the Evaluation and Control of Lead-Based Paint Hazards; the EPA Guidance on Residential Lead-Based Paint, Lead-Contaminated Dust and Lead-Contaminated Soil; the EPA Residential Sampling for Lead: Protocols for Dust and Soil Sampling (EPA Report Number 7474-R-95-001); regulations, guidance methods or protocols issued by States and Indian Tribes authorized by EPA; and other equivalent methods and guidelines.

J. State and Indian Tribal Programs (1996)

The regulations on State and Indian Tribal Programs establish procedures that must be followed by States and Indian Tribes seeking to administer and enforce standards, regulations or other requirements established under Sections 402 or 406, or both, of TSCA.[28] They must apply to the EPA for approval of their programs which must be at least as protective or of human health and the environment as the EPA's program and must provide for adequate enforcement.

State and Tribal applicants may use the EPA's activities regulations as a model state program. As to enforcement, an application must show that State or Tribal officials have (1) authority to enter premises or facilities to inspect for violations; (2) flexible remedies; (3) adequate fiscal and personnel resources; and (4) performance elements, including training of enforcement personnel and inspectors, compliance assistance to the public and the regulated community, facilities and equipment to perform sampling and laboratory analysis, ability to

[27] 40 CFR § § 745.227.

[28] 40 CFR § 745.320 et seq.

process and react to tips and complaints or other information indicating a violation, targeting of inspections, follow-up to inspection reports, compliance monitoring and enforcement to ensure correction of violations, and summary or progress and performance reports to the EPA.[29]

K. Proposed Renovation, Repair and Painting Program Amendments (2006)[30]

On January 10, 2006, the EPA published proposed amendments to the residential property renovation regulations that would significantly expand their coverage. The proposal also included proposed conforming amendments to the activities regulations and the state and tribal program regulations.

The proposed amendments would establish requirements for (1) training renovators and dust sampling technicians; (2) certifying renovators, dust sampling technicians and renovation firms; (3) accrediting providers of renovation and dust sampling technician training; and (4) renovation work practices. The requirements would apply in target housing except housing for the elderly and persons with disabilities (unless a child under age 6 resides or is expected to reside in such housing) and except any 0-bedroom dwelling.

The proposal regulations would take effect in two phases. In the first phase, they would apply to all renovations for compensation in target housing were a child with an increased blood lead level resides, rental target housing built before 1960, and owner-occupied target housing built before 1960 unless the person performing the renovation obtains a statement signed by the owner-occupant that the renovation will occur in the owner's residence and no child under age 6 resides there. In the second phase, they would be extended to all rental target housing and to owner occupied target housing built in the years 1960 through 1977 where a child under age 6 resides. The proposed amendments would allow States, Territories and Indian Tribes to apply for authorization to administer and enforce the new renovation provisions.

The amendments would define a "renovator" briefly to mean a person who either performs or directs uncertified workers who perform renovations. The present definition of

---

29 40 CFR § 745.327.

30 71 Fed. Reg. 1588 (Jan. 10, 2006).

"renovation" would remain the same. The EPA discussed the term "renovation" in the explanation of the proposed regulations at 71 Fed. Reg. 1602.

In its explanation of the proposed amendments, the EPA refers to the existing residential property renovation regulations as the "Pre-Renovation Education Rule." In general, the present information distribution, acknowledgement and certification, and recordkeeping requirements remain in effect with minor modifications. The EPA decided to incorporate the new training, certification, accreditation and work practice requirements in the Pre-Renovation Education Rule (40 CFR § 745.80 et seq.) rather than in the lead-based paint activities regulations (40 CFR § 745.220 et seq.).

A copy of the draft of the EPA's new pamphlet "Protect Your Family From Lead During Renovation, Repair & Painting" is available on the EPA's website.

The EPA announced in the Federal Register, Vol. 72, page 12582, March 16, 2007 the availability of a new study by the EPA titled "Characterization of Dust Lead Levels after Renovation, Repair and Painting Activities" and a new study by the National Association of Home Builders titled "Lead-Safe Work Practices Survey Project."

Protect Your Family From Lead in Your Home

EPA
United States
Environmental Protection
Agency

United States Consumer
Product Safety Commission

U.S. EPA Washington DC 20460
U.S. CPSC Washington DC 20207

EPA747-K-94-001
May 1995

Are You Planning To Buy, Rent, or Renovate a Home Built Before 1978?

Many houses and apartments built before 1978 have paint that contains lead (called lead-based paint). Lead from paint, chips, and dust can pose serious health hazards if not taken care of properly.

By 1996, federal law will require that individuals receive certain information before renting, buying, or renovating pre-1978 housing:

LANDLORDS will have to disclose known information on lead-based paint hazards before leases take effect. Leases will include a federal form about lead-based paint.

SELLERS will have to disclose known information on lead-based paint hazards before selling a house. Sales contracts will include a federal form about lead-based paint in the building. Buyers will have up to 10 days to check for lead hazards.

RENOVATORS will have to give you this pamphlet before starting work.

IF YOU WANT MORE INFORMATION on these requirements, call the National Lead Information Clearinghouse at **1-800-424-LEAD**.

IMPORTANT!

Lead From Paint, Dust, and Soil Can Be Dangerous If Not Managed Properly

FACT: Lead exposure can harm young children and babies even before they are born.

FACT: Even children that seem healthy can have high levels of lead in their bodies.

FACT: People can get lead in their bodies by breathing or swallowing lead dust, or by eating soil or paint chips with lead in them.

FACT: People have many options for reducing lead hazards. In most cases, lead-based paint that is in good condition is not a hazard.

FACT: Removing lead-based paint improperly can increase the danger to your family.

If you think your home might have lead hazards, read this pamphlet to learn some simple steps to protect your family.

Lead Gets in the Body in Many Ways

1 out of every 11 children in the United States has dangerous levels of lead in the blood-stream.

Even children who appear healthy can have dangerous levels of lead.

People can get lead in their body if they:

- Put their hands or other objects covered with lead dust in their mouths.
- Eat paint chips or soil that contains lead.
- Breathe in lead dust (especially during renovations that disturb painted surfaces).

Lead is even more dangerous to children than adults because:

- Babies and young children often put their hands and other objects in their mouths. These objects can have lead dust on them.
- Children's growing bodies absorb more lead.
- Children's brains and nervous systems are more sensitive to the damaging effects of lead.

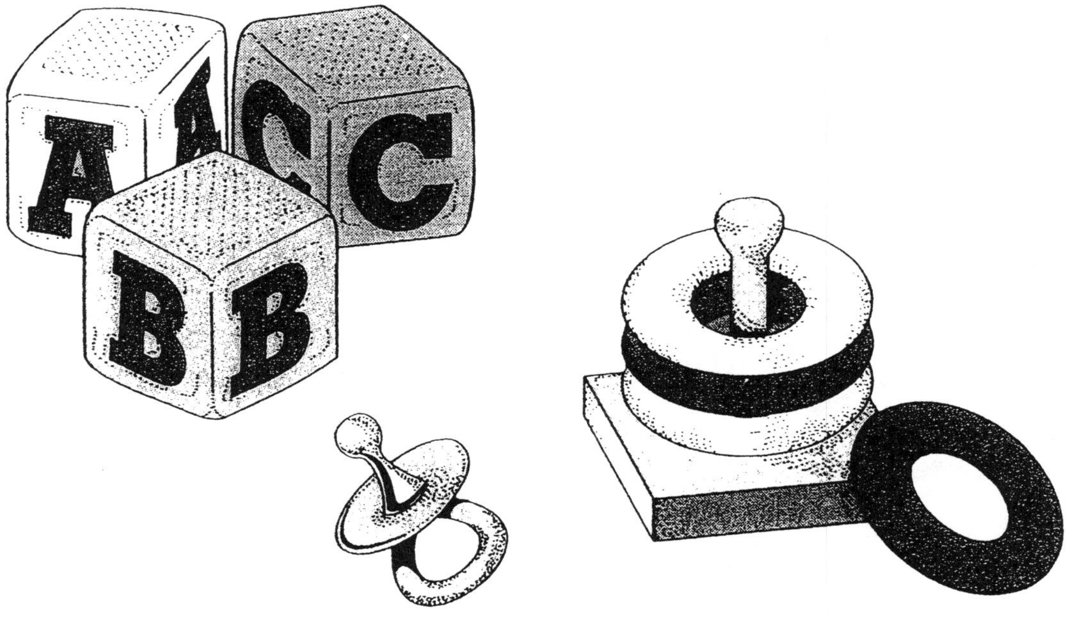

Lead's Effects

If not detected early, children with high levels of lead in their bodies can suffer from:

- Damage to the brain and nervous system
- Behavior and learning problems (such as hyperactivity)
- Slowed growth
- Hearing problems
- Headaches

Lead is also harmful to adults. Adults can suffer from:

- Difficulties during pregnancy
- Other reproductive problems (in both men and women)
- High blood pressure
- Digestive problems
- Nerve disorders
- Memory and concentration problems
- Muscle and joint pain

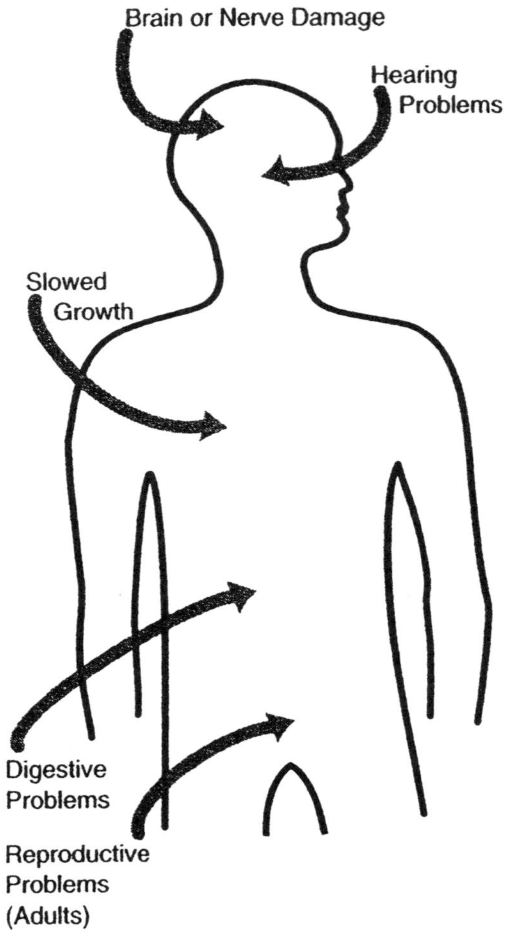

Lead affects the body in many ways.

Checking Your Family for Lead

Get your children tested if you think your home has high levels of lead.

A simple blood test can detect high levels of lead. Blood tests are important for:

- Children who are 6 months to 1 year old (6 months if you live in an older home with cracking or peeling paint).
- Family members that you think might have high levels of lead.

If your child is older than 1 year, talk to your doctor about whether your child needs testing.

Your doctor or health center can do blood tests. They are inexpensive and sometimes free. Your doctor will explain what the test results mean. *Treatment can range from changes in your diet to medication or a hospital stay.*

Where Lead-Based Paint Is Found

In general, the older your home, the more likely it has lead-based paint.

Many homes built before 1978 have lead-based paint. The federal government banned lead-based paint from housing in 1978. Some states stopped its use even earlier. Lead can be found:

- In homes in the city, country, or suburbs.
- In apartments, single-family homes, and both private and public housing.
- Inside *and* outside of the house.
- In soil around a home. (Soil can pick up lead from exterior paint, or other sources such as past use of leaded gas in cars.)

Where Lead Is Likely To Be a Hazard

Lead-based paint that is in good condition is usually not a hazard.

Peeling, chipping, chalking, or cracking lead-based paint is a hazard and needs immediate attention.

Lead-based paint may also be a hazard when found on surfaces that children can chew or that get a lot of wear-and-tear. These areas include:

- Windows and window sills.
- Doors and door frames.
- Stairs, railings, and banisters.
- Porches and fences.

Lead from paint chips, which you can see, and lead dust, which you can't always see, can both be serious hazards

Lead dust can form when lead-based paint is dry scraped, dry sanded, or heated. Dust also forms when painted surfaces bump or rub together. Lead chips and dust can get on surfaces and objects that people touch. Settled lead dust can reenter the air when people vacuum, sweep, or walk through it.

Lead in soil can be a hazard when children play in bare soil or when people bring soil into the house on their shoes. Call your state agency (see page 12) to find out about soil testing for lead.

Checking Your Home for Lead Hazards

Just knowing that a home has lead-based paint may not tell you if there is a hazard.

You can get your home checked for lead hazards in one of two ways, or both:

- A paint **inspection** tells you the lead content of every painted surface in your home. It won't tell you whether the paint is a hazard or how you should deal with it.
- A **risk assessment** tells you if there are any sources of serious lead exposure (such as peeling paint and lead dust). It also tells you what actions to take to address these hazards.

Have qualified professionals do the work. *The federal government is writing standards for inspectors and risk assessors. Some states might already have standards in place.* Call your state agency for help with locating qualified professionals in your area (see page 12).

Trained professionals use a range of methods when checking your home, including:

- Visual inspection of paint condition and location.
- Lab tests of paint samples.
- Surface dust tests.
- A portable x-ray fluorescence machine.

Home test kits for lead are available, but recent studies suggest that they are not always accurate. Consumers should not rely on these tests before doing renovations or to assure safety.

What You Can Do Now To Protect Your Family

If you suspect that your house has lead hazards, you can take some immediate steps to reduce your family's risk:

- **If you rent, notify your landlord of peeling or chipping paint.**
- **Clean up paint chips immediately.**
- **Clean floors, window frames, window sills, and other surfaces weekly.** Use a mop or sponge with warm water and a general all-purpose cleaner or a cleaner made specifically for lead. REMEMBER: NEVER MIX AMMONIA AND BLEACH PRODUCTS TOGETHER SINCE THEY CAN FORM A DANGEROUS GAS.

- **Thoroughly rinse sponges and mop heads after cleaning dirty or dusty areas.**
- **Wash children's hands often, especially before they eat and before nap time and bed time.**
- **Keep play areas clean.** Wash bottles, pacifiers, toys, and stuffed animals regularly.
- **Keep children from chewing window sills or other painted surfaces.**
- **Clean or remove shoes before entering your home to avoid tracking in lead from soil.**
- **Make sure children eat nutritious, low-fat meals high in iron and calcium,** such as spinach and low-fat dairy products. Children with good diets absorb less lead.

How To Significantly Reduce Lead Hazards

Removing lead improperly can increase the hazard to your family by spreading even more lead dust around the house.

Always use a professional who is trained to remove lead hazards safely.

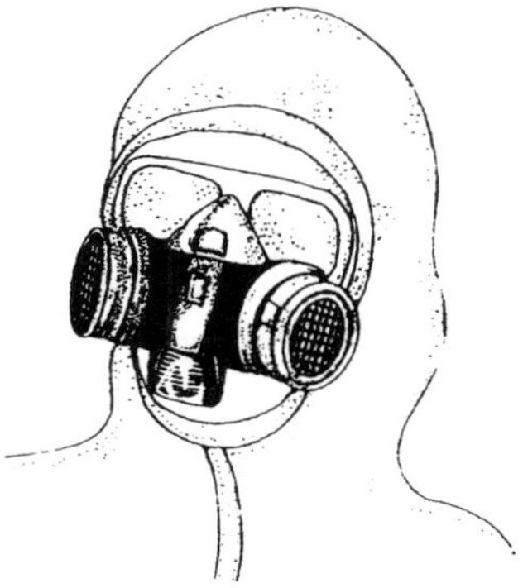

In addition to day-to-day cleaning and good nutrition:

- You can **temporarily** reduce lead hazards by taking actions such as repairing damaged painted surfaces and planting grass to cover soil with high lead levels. These actions (called "interim controls") are not permanent solutions and will need ongoing attention.
- To **permanently** remove lead hazards, you must hire a lead "abatement" contractor. Abatement (or permanent hazard elimination) methods include removing, sealing, or enclosing lead-based paint with special materials. Just painting over the hazard with regular paint is not enough.

Always hire a person with special training for correcting lead problems—someone who knows how to do this work safely and has the proper equipment to clean up thoroughly. If possible, hire a certified lead abatement contractor. Certified contractors will employ qualified workers and follow strict safety rules as set by their state or by the federal government.

Call your state agency (see page 12) for help with locating qualified contractors in your area and to see if financial assistance is available.

Remodeling or Renovating a Home With Lead-Based Paint

Take precautions before you begin remodeling or renovations that disturb painted surfaces (such as scraping off paint or tearing out walls):

- **Have the area tested for lead-based paint.**
- **Do not use a dry scraper, belt-sander, propane torch, or heat gun** to remove lead-based paint. These actions create large amounts of lead dust and fumes. Lead dust can remain in your home long after the work is done.
- **Temporarily move your family** (especially children and pregnant women) out of the apartment or house until the work is done and the area is properly cleaned. If you can't move your family, at least completely seal off the work area.
- **Follow other safety measures to reduce lead hazards.** You can find out about other safety measures by calling 1-800-424-LEAD. Ask for the brochure "Reducing Lead Hazards When Remodeling Your Home." This brochure explains what to do before, during, and after renovations.

If not conducted properly, certain types of renovations can release lead from paint and dust into the air.

If you have already completed renovations or remodeling that could have released lead-based paint or dust, get your young children tested and follow the steps outlined on page 7 of this brochure.

Other Sources of Lead

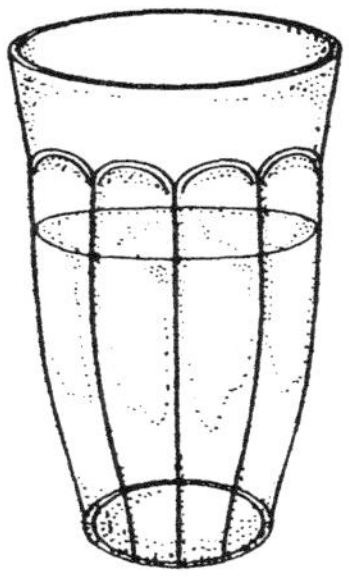

While paint, dust, and soil are the most common lead hazards, other lead sources also exist.

- **Drinking water.** Your home might have plumbing with lead or lead solder. Call your local health department or water supplier to find out about testing your water. You cannot see, smell, or taste lead, and boiling your water will not get rid of lead. If you think your plumbing might have lead in it:
  - Use only cold water for drinking and cooking.
  - Run water for 15 to 30 seconds before drinking it, especially if you have not used your water for a few hours.
- **The job.** If you work with lead, you could bring it home on your hands or clothes. Shower and change clothes before coming home. Launder your clothes separately from the rest of your family's.
- Old painted **toys** and **furniture.**
- Food and liquids stored in **lead crystal** or **lead-glazed pottery or porcelain.**
- **Lead smelters** or other industries that release lead into the air.
- **Hobbies** that use lead, such as making pottery or stained glass, or refinishing furniture.
- **Folk remedies** that contain lead, such as "greta" and "azarcon" used to treat an upset stomach.

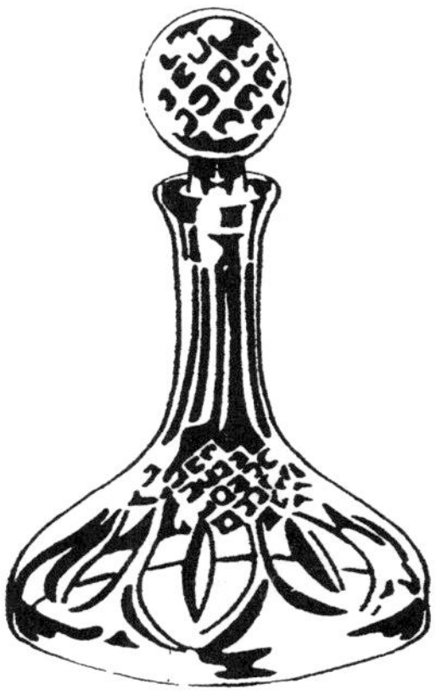

For More Information

The National Lead Information Center

Call **1-800-LEAD-FYI** to learn how to protect children from lead poisoning.

For other information on lead hazards, call the center's clearinghouse at **1-800-424-LEAD.** For the hearing impaired, call, **TDD 1-800-526-5456** (FAX: **202-659-1192,** Internet: **EHC@CAIS.COM**).

EPA's Safe Drinking Water Hotline

Call **1-800-426-4791** for information about lead in drinking water.

Consumer Product Safety Commission Hotline

To request information on lead in consumer products, or to report an unsafe consumer product or a product-related injury call **1-800-638-2772**. (Internet: info@cpsc.gov). For the hearing impaired, call **TDD 1-800-638-8270**.

Local Sources of Information

State Health and Environmental Agencies

Some cities and states have their own rules for lead-based paint activities. Check with your state agency (listed below) to see if state or local laws apply to you. Most state agencies can also provide information on finding a lead abatement firm in your area, and on possible sources of financial aid for reducing lead hazards.

State/Region	Phone Number
Alabama	(205) 242-5661
Alaska	(907) 465-5152
Arkansas	(501) 661-2534
Arizona	(602) 542-7307
California	(510) 450-2424
Colorado	(303) 692-3012
Connecticut	(203) 566-5808
Washington, DC	(202) 727-9850
Delaware	(302) 739-4735
Florida	(904) 488-3385
Georgia	(404) 657-6514
Hawaii	(808) 832-5860
Idaho	(208) 332-5544
Illinois	(800) 545-2200
Indiana	(317) 382-6662
Iowa	(800) 972-2026
Kansas	(913) 296-0189
Kentucky	(502) 564-2154
Louisiana	(504) 765-0219
Massachusetts	(800) 532-9571
Maryland	(410) 631-3859
Maine	(207) 287-4311
Michigan	(517) 335-8885
Minnesota	(612) 627-5498
Mississippi	(601) 960-7463
Missouri	(314) 526-4911
Montana	(406) 444-3671
Nebraska	(402) 471-2451
Nevada	(702) 687-6615
New Hampshire	(603) 271-4507
New Jersey	(609) 633-2043
New Mexico	(505) 841-8024
New York	(800) 458-1158
North Carolina	(919) 715-3293
North Dakota	(701) 328-5188
Ohio	(614) 466-1450
Oklahoma	(405) 271-5220
Oregon	(503) 248-5240
Pennsylvania	(717) 782-2884
Rhode Island	(401) 277-3424
South Carolina	(803) 935-7945
South Dakota	(605) 773-3153
Tennessee	(615) 741-5683
Texas	(512) 834-6600
Utah	(801) 536-4000
Vermont	(802) 863-7231
Virginia	(800) 523-4019
Washington	(206) 753-2556
West Virginia	(304) 558-2981
Wisconsin	(608) 266-5885
Wyoming	(307) 777-7391

EPA Regional Offices

Your Regional EPA Office can provide further information regarding regulations and lead protection programs.

EPA Regional Offices

Region 1 (Connecticut, Massachusetts, Maine, New Hampshire, Rhode Island, Vermont)
John F. Kennedy Federal Building
One Congress Street
Boston, MA 02203
(617) 565-3420

Region 2 (New Jersey, New York, Puerto Rico, Virgin Islands)
Building 5
2890 Woodbridge Avenue
Edison, NJ 08837-3679
(908) 321-6671

Region 3 (Delaware, Washington DC, Maryland, Pennsylvania, Virginia, West Virginia)
841 Chestnut Building
Philadelphia, PA 19107
(215) 597-9800

Region 4 (Alabama, Florida, Georgia, Kentucky, Mississippi, North Carolina, South Carolina, Tennessee)
345 Courtland Street, NE
Atlanta, GA 30365
(404) 347-4727

Region 5 (Illinois, Indiana, Michigan, Minnesota, Ohio, Wisconsin)
77 West Jackson Boulevard
Chicago, IL 60604-3590
(312) 886-6003

Region 6 (Arkansas, Louisiana, New Mexico, Oklahoma, Texas)
First Interstate Bank Tower
1445 Ross Avenue, 12th Floor, Suite 1200
Dallas, TX 75202-2733
(214) 665-7244

Region 7 (Iowa, Kansas, Missouri, Nebraska)
726 Minnesota Avenue
Kansas City, KS 66101
(913) 551-7020

Region 8 (Colorado, Montana, North Dakota, South Dakota, Utah, Wyoming)
999 18th Street, Suite 500
Denver, CO 80202-2405
(303) 293-1603

Region 9 (Arizona, California, Hawaii, Nevada)
75 Hawthorne Street
San Francisco, CA 94105
(415) 744-1124

Region 10 (Idaho, Oregon, Washington, Alaska)
1200 Sixth Avenue
Seattle, WA 98101
(206) 553-1200

CPSC Regional Offices

Eastern Regional Center
6 World Trade Center
Vesey Street, Room 350
New York, NY 10048
(212) 466-1612

Central Regional Center
230 South Dearborn Street
Room 2944
Chicago, IL 60604-1601
(312) 353-8260

Western Regional Center
600 Harrison Street, Room 245
San Francisco, CA 94107
(415) 744-2966

XVIII. THE ATOMIC ENERGY ACT OF 1954

A. Statutory Provisions

The Atomic Energy Act of 1954 (AEA) directs the Nuclear Regulatory Commission (NRC) to license and regulate the production and use of radioactive materials.[1] The AEA also authorizes the NRC to enter into agreements with the states to license and regulate low level radioactive materials.[2] States can also regulate radioactive materials at levels below the level of regulatory concern of the NRC.[3]

B. NRC's Radiation Protection Standards

The NRC has adopted standards for protection against radiation including occupational and public dose limits; control of exposure; respiratory protection and other controls for restricted areas; storage requirements; warning signs, labels and other precautionary requirements; waste disposal requirements; and recordkeeping and reporting requirements.[4] The NRC's radiation protection standards set limits for radioactive lead isotopes. The limits include occupational values (oral ingestion and inhalation); effluent concentrations for direct discharges to air and water; and monthly average concentrations for releases to sewers.[5] The regulations also set quantities (in picocuries) of lead isotopes which require labelling.[6]

C. NRC's Packaging and Transportation Regulations

The NRC has also adopted regulations governing packaging and transportation of radioactive material.[7] The NRC's regulations are coordinated with the DOT regulations described in Chapter XV. The NRC's regulations prescribe packaging specifications and tests; operating controls and procedures; and quality assurance requirements. The regulations apply to

---

[1] 42 USC § 2011 et seq.

[2] 42 U.S.C. § 2021(b).

[3] 42 U.S.C. § 2023.

[4] 10 CFR Part 20.

[5] 10 CFR Part 20, App. B. See List of Elements – Lead is listed as Atomic No. 82. Lead compounds are shown after Atomic No. 82 in Tables 1, 2 and 3.

[6] 10 CFR Part 20, App. C.

[7] 10 CFR Part 71.

packaging and transportation of radioactive lead isotopes. For example, activity values, designated as A_1 and A_2, are prescribed for certain radioactive lead isotopes.[8]

D. State Regulation

In agreement states, such as New York, state agencies adopt and administer regulations similar to those of the NRC. For example, the New York State Department of Health licenses and regulates the use of radioactive materials and equipment by hospitals, except for those regulated by the City of New York Department of Health. The New York State Department of Labor licenses and regulates the use of radioactive materials and equipment by industry. Their regulations also apply to any radioactive lead isotopes that may be used by their licensees.

---

[8] 10 CFR Part 71, App. A.

XIX. STATE ENVIRONMENTAL AND HEALTH LAWS

A. State Environmental Laws

The states have adopted environmental laws implementing the federal environmental laws including the CAA, CWA, SWDA, RCRA and CERCLA. These state environmental laws regulate lead and lead compounds in a manner similar to the federal environmental laws described in earlier chapters. The reason is that the federal laws require that the state laws be at least as protective of human health and the environment as the federal laws and that the state laws and regulations be approved by the EPA. Indeed, many state environmental agencies have incorporated large portions of the EPA's regulations by reference into their regulations.

The regulations of the state environmental agencies regulate air emissions, water discharges, drinking water content, waste disposal and cleanup locations at hazardous waste facilities involving lead and lead compounds by standards which are the same as, or stricter than, those of the EPA. Examples of state laws going beyond the federal laws include the New Jersey Industrial Site Recovery Act;[1] the Connecticut Transfer Act;[2] the Illinois Responsible Property Transfer Act;[3] and the Indiana Environmental and Hazardous Disclosure and Responsible Property Transfer Law, 13 Indiana Code.[4] Such laws may require cleanup of lead contamination in soil and groundwater upon transfers of industrial and commercial, even though the CWA, RCRA and CERCLA may not require a cleanup.

B. State Health and Housing Laws

Many states have also adopted health or housing laws to regulate lead-based paint hazards. Some of these laws have established only limited programs. Others provide for comprehensive programs which may be found in the laws and in the implementing regulations

---

[1] 13 NJSA § 13:1K-6 et seq.

[2] 22a Conn. Gen. Stat. Annot., § 22a-134 et seq.

[3] 765 ILCS, § 90/1 et seq.

[4] § 13-7-22.5-1 et seq.

adopted by health, environmental, and housing agencies. In general, the more comprehensive laws and regulations are found in the northeastern states.

Examples of the provisions of the comprehensive programs are screening of children; mandatory reports of "lead poisoning" as defined in terms of micrograms per deciliter (ug/dL) of blood; inspection and evaluation of residences for lead-based paint as defined in terms of lead content by weight or per square centimeter; abatement of lead-based paint hazards as specified; notice to a designated agency prior to abatement work; licensing and training of abatement contractors and personnel; prohibition of retaliation by owners against tenants; and imposition of strict liability on owners under some circumstances.

The regulations in states such as Massachusetts and Maryland are quite detailed, particularly as to the methods, equipment and performance standards for abatement. For example, abatement rules often prescribe inspection methods and equipment (x-ray fluorescence analyzer, atomic absorption analyzer, or sodium sulfide); occupant and property removal or protection; methods for containment of the work area with polyethylene sheeting and sealants; warnings and restriction of access to the work area; the surfaces which must be abated; prohibited abatement methods; permitted abatement methods; cover materials useable for encapsulation; worker protection during abatement; cleanup requirements such as HEPA filters; reoccupancy inspection requirements; lead dust levels during and at the end of abatement work; disposal of wastes; recordkeeping; and issuance of certificates of compliance.

All state laws and regulations governing lead-based paint activities were required to conform to standards set by the EPA within two years after the EPA issued its final regulations on lead-based paint activities under Subchapter IV of TSCA. The EPA's regulations are described in Chapter XVII (I). The EPA announced that its regulations should be treated as a model state program, but also outlined the major program elements that state programs must have if they prefer their own programs.

Each state seeking to administer and enforce standards, regulations and other requirements for lead-based paint activities must apply to the EPA for approval of its program which must be at least as protective of human health and the environment as the EPA's program.

Each state program must also provide for adequate enforcement. If any state did not have an approved program within two years after the EPA's final regulations are published, TSCA requires the EPA to establish a federal program in the state.

References to many of the existing state health and housing laws relating to lead-based paint are provided in Appendix A at the end of this Chapter.

C. State Laws Restricting Lead in Water Pipe and Solder

Many states have laws restricting the use of piping and solder containing lead in community water systems. References to many of these laws are provided in Appendix A at the end of this Chapter.

D. State Lead-Acid Battery Recycling Laws

Many states have adopted statutory provisions requiring the recycling of lead acid batteries and other batteries. References to many of these laws are provided in Appendix A to this Chapter. Other states accomplish recycling under their general environmental laws and regulations. A summary of U.S. State Lead-Acid Battery Laws by Battery Council International is Appendix B to this Chapter.

E. State Laws Restricting Lead in Packaging

In 1994, Pennsylvania adopted the Safe Packaging Act.[5] The Act prohibits the sale or offer for sale by a manufacturer or distributor in Pennsylvania of any package or packaging component which includes inks, dyes, pigments, adhesive, stabilizers or any other additives containing lead or certain other metals which has been intentionally introduced as an element during manufacturing or distribution, as opposed to its incident presence. The Act also prohibits the sale or offer for sale of any product in such a package or packaging. The concentration level of lead in any package or packaging may not exceed 600 ppm (0.06%) by weight.[6] The Act contains several exemptions including exemptions related to recycling programs.[7]

---

[5] 35 P.S. § 6024.101 et seq.

[6] Id., § 6024.301.

[7] Id., § 6024.302.

Manufacturers and supplies are required to certify compliance.[8]

Other states have also adopted safe packaging laws. For example, as shown in the List of State Laws, which is Appendix A to this Chapter, safe packaging laws have been adopted in Illinois, Washington and Wisconsin.

F. The City of New York; A Legislative History

A number of municipalities, especially in the northeastern states, have adopted local laws on lead-based paint that have increased in strictness and detailed requirements as the years passed.

The experience in the City of New York is illustrative. Local Law No. 1 of 1982 required the owner of a multiple dwelling to remove or cover lead-based paint in any dwelling unit in which a child six years and under resided in a manner approved by the housing department. Peeling paint in any multiple dwelling erected before 1960 was presumed to be lead-based paint. The Health Code authorized the health department to order abatement of lead-based paint in housing occupied by a lead-poisoned child. However, since owners of multiple dwellings usually could determine whether they contained lead-based paint only by conducting costly inspections, owners generally did not act unless ordered to do so after a peeling paint inspection by the building department or discovery of a child with an elevated blood lead level by the health department through the school "finger stick" testing program.

To sponsor more active compliance, the City enacted Local Law 38 in 1999, requiring the owner of a multiple dwelling to correct all lead-based paint hazards. It created an affirmative obligation to abate lead-based paint in deteriorated condition and in dwellings where a child with an elevated blood lead concentration resided. The author presented information to the City Council in favor of the new law. However, Local Law 38 was opposed by advocacy groups who wanted a law requiring more extensive and detailed requirements. They succeeded in having the new law declared invalid on grounds of insufficient compliance with the

[8] Id., § 6024.303.

procedures of the State Environmental Quality Review Act. New York City Coalition v. Vallone, 100 N.Y.2d 337, 794 N.E.2d 672 (2003).

The City then adopted Local Law 1 of 2004. This new law requires the owner of a multiple dwelling to prevent the reasonably foreseeable occurrence of lead-based paint hazards and to remediate them and underlying defects using work practices prescribed by the health department. The law creates a rebutable presumption that paint in a pre-1960 multiple dwelling where a child under age 7 resides or in the common areas is lead-based paint. The owner of pre-1960 multiple dwellings, and owners of 1960 through pre-1978 multiple dwellings who have actual knowledge of the presence of lead-based paint, must investigate dwelling units where any child under age 7 resides and common areas for peeling paint chewable surfaces, deteriorated subsurfaces, friction surfaces and impact surfaces. The investigation must be undertaken at least once a year and more often if the owner knows or should know of a condition that is likely to cause a lead paint hazard, if an occupant complains, or if the housing department issues a notice of violation or issues a corrective order. Occupants are required to provide accurate and truthful information about residency of a child under age 7 and to provide reasonable access to the owner to investigate and repair. Owners must inform occupants of the results of investigations and furnish copies of reports to them. Reports must be kept for ten years.

Owners of multiple dwellings must provide to the occupant of a dwelling unit upon signing of a lease, lease renewal or agreement to lease, or upon occupancy if there is no lease, a notice in English and Spanish inquiring whether a child under 7 resides or will reside in the unit. Therefore, the notice must be given annually by January 16 of each year. An owner who does not receive a response must take further actions, including notice to the health department.

Upon turnover of any dwelling unit in a pre-1960 multiple dwelling or a dwelling unit in a pre-1960 private dwelling to be occupied by persons other than the owner or the owner's family, the owner must within the dwelling unit: (1) remediate all lead-based paint hazards; (2) make all bare floors, window sills and window wells smooth and cleanable; (3) provide for removal or permanent covering of all lead-based paint on all friction surfaces on all doors and door frames; and (4) provide for removal or permanent covering of all lead-based paint on all friction surfaces on all windows or provide for installation of replacement window channels or slides on all lead-based painted friction surfaces on all windows.

The law provides for inspections and enforcement by the building department and requires that the department provide employee training in the use of XRF equipment and other methods and adopt regulations on compliance procedures. The law also provides for inspections and enforcement by the health department.

Among its numerous additional requirements, the law requires that the building department adopt regulations prescribing work practices for work done to remove violations and work done when no violation has been issued. The department has published a Guide to Local Law 1 of 2004 Work Practices. The law also requires the health department to investigate to identify potential sources of elevated blood lead level if there is a report of a person under age 18 with a blood lead level of 15 ug/dL or higher residing in a dwelling unit. The health department must prepare and distribute annually a brochure advising medical providers about the law on screening and testing children. The health department must also prepare and distribute a pamphlet on lead poisoning prevention. The law contains numerous provisions imposing civil and criminal penalties for failure to meet its requirements.

The law also applies to day care services that regularly give care to seven or more children under age six for more than five hours per week or one month per year.

The law does not apply to multiple dwelling units where title is held by a cooperative or condominium and the unit is occupied by the shareholder of record or his or her family. For several years, real estate developers have been demolishing or "gut" renovating many old buildings in the City of New York and converting the properties to cooperative and condominium housing for affluent customers. This activity should reduce the number of buildings containing lead-based paint, but will also reduce the supply of affordable housing.

Appendix A

LIST OF STATE LAWS

The following list of state laws was prepared by surveying the indices of the state statutes annotated for lead and topics closely related to lead such as lead-based paint and batteries. Readers should keep in mind that index preparers do not always identify subjects covered by multipurpose laws. The programs in some states are contained extensively in the regulations of health, environmental or housing agencies.

Alabama: Code of Alabama, Alabama Lead Ban Act of 1988, § 22-37-1 et seq. (Prohibits materials which are not lead free in drinking water systems or plumbing); Alabama Lead Reduction Act of 1997 § 22-37A-1 et seq. (Regulates persons engaged in lead hazard reduction activities).

Arizona: Arizona Revised Statutes, § 36-1671 et seq. (Lead-Based Paint); § 41-2121 (Oxygenated Fuel); § 44-1321 et seq. (Sale and Disposal of Batteries); § 44-1641 (Scrap Metal Dealers); § 49-353 (Pipes, Fittings, Solder, Flux).

Arkansas: 20 Arkansas Code, § 20-27-601 (Lead Poisoning Prevention); Lead-Based Paint Hazard Act § 8-4-401 (Training and licensing of lead-based paint activities).

California: Health and Safety Code, § 12000 et seq. (High Explosives – Lead Azide); § 17920.10 (Lead Hazards in Buildings); § 105185 et seq. (Occupational Lead Poisoning Prevention); § 105185 et seq. (Accreditation of Lead-Based Paint Training Providers); § 105275 et seq. (Childhood Lead Poisoning Prevention Act); § 108550 et seq. (Toy Safety); § 108850 et seq. (Product Safety – Control of Lead Release from Tableware); § 110552 et seq. (Food, Drug and Cosmetics – Adulterated Candy); § 116875 et seq. (Lead free pipes, plumbing fittings and fixtures, and solder and flux); Civil Code.§ 2079 (Duty to Disclose Environmental and Other Hazards to Purchasers of Residential Property); § 1102 (Disclosures Upon Transfer of Residential Property); Business and Professions Code. § 10084.1 (Consumer Education Booklet); Public Resources Code. § 42440 et seq. (Recycled Battery Programs), § 42451 et seq.

(Rechargeable Battery Recycling Act of 2006); Education Code § 32240 et seq. (Lead Safe Schools Protection Act); Safe Drinking Water and Toxic Enforcement Act of 1986 (Proposition 65), Office of Environmental Health Hazard Assessment.

Colorado: Colorado Revised Statutes, § 25-5-501 et seq. (Hazardous Substances Act); § 25-5-1101 et seq. (Prevention, Intervention, and Reduction of Lead Exposure); § 25-7-1101 et seq. (Lead-Based Paint Abatement).

Connecticut: Connecticut General Statutes Annotated, § 19a-110 et seq. (Lead poisoning; reports, treatment centers, investigation, prevention, education and abatement); Department of Health Services Regulation § 19a-111 et seq.; § 8-219e (Financial assistance for removal of lead-based paint); § 10-206 and 206b (Health assessments); § 20-474 et seq (Lead Abatement Consultants, Contractors and Workers); § 22a-256g et seq. (Lead acid and nickel-cadmium batteries); § 25-39e (Restriction on use of lead solder in potable water systems); § 17b-808 (Emergency housing).

Delaware: Delaware Code Annotated, Health and Safety, Title 16, §122 (Powers of Department of Health and Social Services); § 2601 et seq. (Childhood Lead Poisoning Prevention Act); Insurance Code, Title 18, § 3337 and § 3554 (Lead poison screening reimbursement by insurers).

District of Columbia: D.C. Code § 7-871-01 et seq. (Childhood Lead Poisoning Screening and Reporting Act); § 8-115.1 et seq. (Lead-Based Paint Abatement and Control); § 34-2151 (Lead Pipe Replacement Assistance); § 42-3251 et seq. (Multiple Dwelling Residence Water Lead Level Testing).

Florida: 14A Florida Stat. Annot. § 403.717 et seq. (Lead-acid batteries)

Georgia: Official Code of Georgia, § 12-8-28 et seq. (Lead Acid Vehicle Batteries); § 31-41-1 et seq. (Lead Poisoning Prevention Act). The effective date of Article 2 of this Act on childhood lead exposure control depended on the receipt of sufficient federal funding which was reportedly received); § 16-7-81 (Explosive Materials).

Hawaii: Hawaii Revised Statutes, § 321.11 (27) (Accreditation of training); § 340E-4.7 (Notice to public water system users of potential lead contamination); § 342I-1 et seq. (Lead-acid batteries); § 342P-1 et seq. (Lead Contamination).

Idaho: Idaho Code, § 39-7001 et seq. (Sale and disposal of lead-acid batteries).

Illinois: 27 Illinois Compiled Statutes Annot., Lead Poisoning Prevention Act, Health and Safety, 410 ILCS 45/1 et seq.; Environmental Safety, 415 ILCS 5/21.5 (Toxic packaging reduction); 415 ILCS 5/22.16b (Municipal waste incinerator permit limits); 415 ILCS 5/22.23 (Batteries); 415 ILCS 5/22.23a (Fluorescent and high intensity discharge lamps); 415 ILCS 505.8 et seq. (Lead minerals).

Indiana: Burns Indiana Stat. § 20-34-3-11 (Lead poisoning tests); § 13-17-1 et seq. Lead-Based Paint Activities); § 16-41-39.4-1 et seq. (Childhood Lead Poisoning).

Iowa: Iowa Code Annotated, § 92.8 (child labor, exposure to lead fumes or compounds); § 455D.10 et seq. (Lead Acid Batteries: Disposal and Recycling); § 135.100 et seq. (Lead Abatement Programs)

Kansas: Kansas Statutes Annot. § 2-2203 (Prohibition on sale of lead arsenate); §65-1,203 (Residential Childhood Lead Poisoning Prevention).

Kentucky: Kentucky Revised Stat. Annot., Vol. 9, § 211.900 et seq. (Lead Poisoning Prevention); § 211.9061 (Lead Hazard Detection and Abatement); § 217.801 (Restrictions on Sale and Use of Lead-Based Paint); § 224.50-410 (Disposal of Lead-Acid Batteries)

Louisiana; West's Louisiana Revised Stat., § 30:2351 et seq. (Lead Hazard Reduction, Licensure and Certification); § 30.2419 et seq. (Lead Acid Batteries); § 40.1299.20 et seq. (Lead Paint Poisoning Prevention and Control Act); § 40.733 (Bottled water)

Maine: Maine Revised Statutes Annot. Ch.22 § 1314 et seq. (Lead Poisoning Control Act); Ch. 33 § 171 et seq. (Residential Property Disclosures); Ch. 12-B §1291 et seq. (Lead Abatement).

Maryland: Environmental Code, 6-301 et seq., (Lead-Based Paint); 6-801 et seq. (Reduction of Lead Risk in Housing); 6-1001 et seq. (Accreditation of Abatement Services); 6-1101 et seq. (Rechargeable Batteries) 7-106 et seq.; Real Property Code §8-211.1 (Rent Escrow)

Health-General § 18-106 (Lead Poisoning Screening Program); Housing and Community Development Department, Lead Hazard Reduction Grant and Loan Programs, § 4-701 et seq.; Insurance, § 19-701 (Lead Hazard Coverage for Rental Property); Real Property, §14-123 (Nuisance Actions by Community Associations).

Department of the Environment Regulations, Title 26, Subtitle 02, Section 26.02.07, Procedures for Abating Lead Containing Substances from Buildings.

Massachusetts: Gen. Laws of Mass., Public Health, Chapter 111, § 111.189.A et seq. (Lead Poisoning Prevention and Control); Ch. 112, § 222 (Licensed Lead Paint Inspectors); Ch. 270, § 1 (Adulteration of liquor with sugar of lead); Banks and Banking, Ch. 167, § 48 (Financing of Premises Containing Lead in Paint, Plaster or Materials); Insurance, Ch. 175, § 47C (Insurance Coverage for Newborn Infants or Adoptive Children).

Department of Public Health, Lead Poisoning Prevention Regulation, 105 CMR 460.00; Deleading Regulation, 454 CMR 22.00; Occupational Lead Poisoning Regulation, 454 CMR 23.000.

Michigan: Michigan Compiled Laws, § 333.5111 (Reports); § 333.5451 (Lead Abatement); § 333.12104 (Lead based paint control program); § 333.20531 (Reporting blood lead test results); § 750.243a et seq. (Fireworks containing lead sulphides); § 125.1513b (Lead-free plumbing systems); § 324.17101 et seq. (Battery Disposal); § 286.341 et seq. (Fruit protection, poisonous spray residue tolerances); § 287.141 et seq. (Livestock and poultry remedies); § 290.641 et seq. (Motor fuels quality).

Minnesota: Minnesota Statutes Annotated. § 119A.46 (Lead Abatement Program); § 144.9501 et seq. (Lead Poisoning Prevention Act); § 115A.915 et seq. (Batteries); § 115A.965 et seq. (Toxics in Packaging); § 116.87 et seq. (Residential lead paint waste); § 299F.19 (Lead azide and lead oxide).

Mississippi: Mississippi Code, § 17-17-429 et seq. (Disposal of lead acid batteries); § 49-17-501 et seq. (Lead-Based Paint Activity Accreditation and Certification Act.); § 75-55-5 (Gasoline and Petroleum Products) § 69-23-5 (Pesticides); § 49-25-1 et seq. (Junkyard Control Law).

Missouri: Vernon's Annotated Missouri Statutes. § 260.820 et seq. (Lead in packaging components); § 260.860 et seq. (Disposal and sale of batteries); § 640.120 et seq. (Lead free materials in water system construction); § 701.300 et seq. (Lead Abatement and Prevention of Lead Poisoning).

Nebraska: Revised Statutes of Nebraska, § 71-2513 et seq. (Childhood Lead Poisoning Prevention Act); § 71-6318 (Residential Lead-Based Paint Professions Certification Act).

Nevada: Nevada Revised Statutes, § 444.350; (No city or county may allow use of solder or flux containing in excess of 0.2% lead or pipe containing in excess of 8% lead in water systems); § 444.583 (Unlawful disposal of motor vehicle battery).

New Hampshire: New Hampshire Revised Statutes Annotated, Vol. 10, Ch. 130A (Lead Paint Poisoning Prevention and Control Act); § 477:4-a (Conveyances of realty and interests therein); Ch. 141A (Critical Health Problems Reporting Act); § 149-M:28 (Restrictions on Battery Sales and Disposal).

New Jersey: 13 N.J. Stat. Annot., Conservation and Development, Dry Cell Battery Management Act, § 13:1E-99.59 et seq.; Recycling of Used Lead Acid Batteries, § 13:1E-199; 24 N.J. Stat. Annot., Food and Drugs, Paint Containing Lead § 24:14A-1 et seq.; 26 N.J. Stat. Annot., Lead Poisoning Abatement and Control, § 26:2-130 et seq. Lead Evaluation and Abatement, § 26:2Q-1 et seq. 52 N.J. Stat. Annot., § 52:27D-437.1 et seq. Lead Hazard Control Assistance Act.

New Mexico: New Mexico Statutes Annotated. § 69-7-6 (Lead Poisoning-occupational).

New York: Public Health Law, Commissioner; general powers and duties, § 206(1)(n) and Control of Lead Poisoning, § 1370 et seq.; Environmental Conservation Law, Battery management and disposal, § 27-0719, Lead-acid battery recycling, § 27-1701, Enforcement, § 71-2722.

State Department of Health Regulations, Part 67, Lead Poisoning and Control.

City of New York, Health Code §§ 179 to 181; Administrative Code and Charter, Title 27, Chapter 2, Housing Maintenance Code, § 2056 et seq., commonly called Local Law No. 1.

City of New York, Rules, Volume 10, Housing, Title 28, Chapter 11, Use of Lead-Based Paint in Multiple Dwellings.

North Carolina: General Statutes of North Carolina, Title 17, § 130A-309.70 et seq. (Lead acid batteries); § 130A-131.5 et seq. (Lead Poisoning in Children).

North Dakota: North Dakota Century Code Annot., § 23-25-03.1 (Lead-Based Paint Licensing and Certification); § 23-29-05.2 (Lead acid batteries).

Ohio: Page's Ohio Revised Code Annot., Title 37, Chapter 3742 (Lead Abatement); Title 53, § 5302.30(D) (Residential real property transfer disclosures).

Oklahoma: Oklahoma Statutes Annotated, 27A, § 2-12-101 (Lead-Based Paint Management Act); 27A, § 2002 (Lead Acid Batteries); 45, § 411 et seq. (Lead and Zinc Mining); 63, § 1-114.1 (Comprehensive Childhood Lead Poisoning Prevention Program); 27A, 2-5-106.1 (Controlled Open Burning).

Oregon. Oregon Revised Statutes, § 431.920 (Training Provider Accreditation) § 459.420 et seq. (Lead Acid Batteries); § 448.330 (Water Pipes and Fittings); 701.500 et seq. (Lead Poisoning and Hazard Reduction).

Pennsylvania: Purdon's Pennsylvania Statutes, Vol. 35, § 723.1 et seq. (Plumbing System Lead Ban and Notification Act); Vol. 35, § 5901 et seq. (Lead Certification Act); Vol. 35, § 6024.101 et seq. (Safe Packaging Act); Vol. 43, § 471 et seq. (Lead Manufacturing Establishments); Vol. 53, § 4000.1510 (Lead acid batteries); Vol. 77, § 27.1 (Occupational diseases); 3 Pa. C.S.A. § 5901 et seq. (Organic Food Products Standard Act).

Rhode Island: General Laws of Rhode Island Annot., § 23-24.6-1 et seq. (Lead Poisoning Prevention Act); § 23-24.7-1 et seq. (Occupational Health Lead Protection Act); § 23-60.1-1 (Dry Cell Battery Control); § 42-11-14 (Prohibited purchase of lead based paint for painting public buildings).

South Carolina: Code of Laws of South Carolina, § 44-53-1310 et seq. (Childhood Lead Poisoning Prevention and Control Act); § 44-96-180 (Lead acid batteries).

South Dakota: South Dakota Codified Laws, Vol. 11B, § 34A-6-91 (Used lead acid batteries).

Tennessee: Tennessee Code Annotated, § 39-17-1401 et seq. (Metals Recycling); § 68-211-866 (Disposal of lead acid batteries); § 68-221-720 (Lead free pipe, solder or flux in public water systems); § 68-131-401 et seq. (Lead-Based Paint Abatement Certification Act).

Texas: Vernon's Texas Statutes and Codes Annot., Health and Safety, § 84.001 et seq. (Occupational diseases); §88.001 (Reports of Childhood Lead Poisoning); § 161.101 (Poisoning tests); § 361.451 (Lead acid batteries); § 361.138 (Definition of hazardous substances and waste); § 372.0035 (Plumbing); § 382.053 (Lead smelting plants, construction permits); Occupations, § 1955.001 et seq. (Lead-Based Paint Abatement).

Utah: Utah Code Annotated, § 19-6-601 (Lead Acid Batteries); § 19-2-104(3) (Air Quality Board Powers).

Vermont: Vermont Statutes Annotated, Title 18, § 1751 et seq. (Lead Poisoning); Title 10, § 400 (Loan guarantees for lead based paint hazard reduction); Title 10, § 6621a (Landfill disposal requirements); Title 9, § 3801 et seq. (Labelling of white lead paint).

Virginia: Code of Virginia, § 10.1-1425.1 et seq. (Lead acid batteries); § 32.1-46.1 (Testing of Children); § 36-106 (Lead hazard control violations); § 36-107.1 (Sale of residential structure with levels exceeding building code standards); § 40.1-51.20 (Duties and inspections of abatement contractors); § 54.1-500 et seq. (Licensing of contractors).

Washington: West's Revised Code, Washington Annotated, § 70.95.610 et seq. (Battery disposal and sales); § 70.95G.005 et seq. (Packaging concentration limits for metals including lead); § 70.103.010 et seq. (Lead-Based Paint – Accreditation and Certification); § 70.140.010 et seq. (Area-Wide Soil Contamination within the central Puget Sound smelter plume).

West Virginia: Michie's West Virginia Code, Public Health, § 16-35-1 (Lead Abatement); § 22-15A-22 (Lead-acid batteries disposal program).

Wisconsin: Wisconsin Statutes Annotated, § 254.11 et seq. (Toxic Substances including lead hazards, lead poisoning and lead hazard reduction); § 287.07 and § 287.18 (Lead acid batteries); § 29.311 (Shotshell restrictions); § 100.285 (Reduction of toxics in packaging); § 48.737 (Day care centers); § 632.895 (Accident and health insurance – mandatory coverage of lead poisoning screening).

Wyoming: Wyoming Statutes Annotated, § 30-1-101 et seq. (Mines and Minerals); § 35-11-509 et seq. (Lead acid batteries).

SUMMARY OF U.S. STATE LEAD-ACID BATTERY LAWS

STATE/COUNTY	EFFECTIVE DATE	BCI MODEL	DEPOSIT[a] (REFUNDABLE)	SPLIT OF DEPOSIT	DEPOSIT REFUND PERIOD	POINT OF SALE SIGN[b]	FEE (NONREFUNDABLE)
Arizona[c]	09/27/90	Yes	$5	100% Retailer	30 days	Retailer	
Arkansas	07/1/92	Yes	$10	100% Retailer	30Days	State	
California	01/1/89	Yes				No	
Connecticut[d]	10/1/90	Yes	$5	100% Retailer	30 days	Retailer	
Florida	01/1/89	Yes				No	$1.50[o]
Georgia	01/1/91	Yes				Retailer	
Hawaii	01/1/90	Yes				State	
Idaho[1]	07/1/91	Yes	$5	100% Retailer	30 days	Retailer	
Illinois	09/1/90	Yes				Retailer	
Indiana	01/1/91	Yes				Retailer	
Iowa	07/1/90	Yes				Retailer	
Kansas City,[e] Missouri	03/14/90	Yes				Retailer	
Kentucky	07/13/90	Yes				Retailer	
Louisiana	09/1/89	Yes				Retailer	
Maine	10/30/89	Yes	$10	100% Retailer	7 days	State	$1.00[p]
Massachusetts	12/31/90	No[g]				No	
Michigan	04/1/90	Yes				State	
Minnesota	10/4/89	Yes[f]	$5	100% Retailer		State	
Mississippi	07/1/91	Yes				State	
Missouri	01/1/91	Yes				State	
Nebraska	09/1/94	No[i]					

Nevada	01/1/92	No[g]				No	
New Hampshire	01/1/91	No[g]				No	
New Jersey	10/9/91	Yes				Retailer	
New Mexico	12/31/91	No[g]					
New York	01/1/91	Yes	$5	100% Retailer	30 Days	Retailer	
North Carolina	01/1/91	Yes				Retailer	
North Dakota	01/1/92	Yes				No	
Oklahoma	09/1/93	Yes[t]				Retailer	
Oregon[h]	01/1/90	Yes				Retailer	
Pennsylvania	07/26/89	Yes				State	
Rhode Island	01/1/89	Yes	See[u]			State	
South Carolina	05/27/91	Yes	$5	100% Retailer	30 days	State	$2[m]
South Dakota[r]	07/1/92	Yes				No	
Tennessee	07/1/90	Yes[q]				No	
Texas	09/1/91	Yes				State	$2/$3[n]
Utah[k]	01/1/92	Yes				Retailer/Wholesaler	
Vermont	06/17/94	Yes				Retailer	
Virginia	07/1/90	Yes				State	
Washington	07/23/89	Yes	$5	100% Retailer	30 days	State	
West Virginia	04/6/94	Yes[s]				Retailer/Wholesaler	
Wisconsin	01/1/91	Yes	See[j]			State	See[j]
Wyoming	06/8/89	Yes				State	

Footnotes:

a This refers to a deposit in lieu of trade-in. The deposit is paid by any person who purchases a new battery without returning a used battery at the same time. (T).

b This refers to whose responsibility it is to make the educational signs, the state or the retailer. A "No" indicates that there is no sign requirement."

c AZ requires all lead batteries sold to be labeled with a universally accepted recycling symbol. AZ also requires that State agencies and political subdivisions comply with the battery recycling law.

d Retailers in CT must take back batteries one-for-one at the point of sale.

e Kansas City's ordinance requires that retailers take back up to 3 batteries not at the point of sale, and it requires that junk batteries be stored in "an adequately ventilated enclosure in good repair that protects its contents from any precipitation, etc." Any spilled acid must be immediately collected and neutralized.

f MN now requires that retailers take back up to 5 batteries not at the point of sale.

NH, NM, NV and MA placed a ban on the landfilling and incineration of lead batteries only. NV will al

g	lead battery disposal at stat "permitted" facilities, however.
h	OR requires that until 12/31/93 retailers must accept at least 1 battery from consumers, after which they must only accept batteries one-for-one at the point of sale.
i	NE placed a prohibition on only the landfilling of lead batteries.
j	WI law allows retailers to charge a $5 deposit in lieu of a trade-in, and to charge $3 for taking a battery.
k	UT requires retailers to take back a maximum of two used lead batteries from customers. In addition to the BCI model law, a 1998 regulation prohibits solid waste disposal of lead acid batteries.
l	ID requires all lead batteries sold to be labeled with a universally accepted recycling symbol. In addition, batteries used in motorcycles, off-road recreation vehicles or lawn and garden equipment are exempt from the deposit in lieu of a trade-in requirement.
m	SC requires retailers to collect a $2.00 fee for lead batteries sold to the ultimate consumer. The retailer may retain three percent of the collected fees to cover administrative costs. Fees collected by the state treasurer are to be deposited into a Solid Waste Management Trust Fund. Small sealed lead-acid batteries are now exempt from the fee and BCI model provisions; however, a study on the recycling of these batteries is required. See S.C. Code Ann. x 44-96-40(23).
n	TX requires the collection of a $2.00 and $3.00 fee for batteries less than 12volts, and, equal to or greater than 12 volts respectively. Exempted from the fee is any battery that is: 1) rated at less than 10 ampere hours; 2) sealed so that no access to the interior of the battery is possible without destroying the battery; and 3) with dimensions (sum of height, width and length) less than 15 inches. The fees are to be collected by any wholesaler or retailer who sells a battery not for resale. To cover administrative costs, the dealer may retain 2-1/2 cents per unit. All remaining money, less four percent to cover state administrative costs, goes to the state comptroller to be placed in a waste remediation fund.
o	FL requires the collection of a $1.50 fee per battery at the retail level.
p	ME requires the collection of a $1.00 fee per battery at the retail level.
q	TN prohibits landfills or incinerators in the state from accepting lead-acid batteries for incineration or disposal. Further, lead-acid battery retailers must accept used lead-acid batteries as trade-in batteries.
r	SD requires wholesalers and retailers to "accept, on a one for one exchange basis, used lead-acid batteries and . . . ensure the proper handling and disposal of the batteries." Further, after July 1, 1995, all lead-acid batteries shall be eliminated from landfilled wastes."
s	WV requires retailers and wholesalers to collect used lead-acid batteries from customers and post point-of-sale signs.
t	OK requires that retailers of lead-acid batteries post and maintain a sign at or near the point of display or sale to inform the public that lead-acid batteries are accepted for recycling.
u	RI law specifies that retailers may voluntarily add a core charge (amount unspecified) to the price of a new vehicle battery. The core charge must be refunded if a used battery is returned within 7 days of the date of purchase.

Final Note: Several states have adopted separate household or dry cell battery recycling laws that include provisions strictly applicable to small sealed lead-acid batteries. These states are California, Florida, Illinois, Iowa, Maine, Maryland, Minnesota, New Hampshire, New Jersey, New York, Oregon, and Vermont.

Lead-Acid Battery Definitions by State

Arizona	"[A] battery with a core of elemental lead and a capacity of six or more volts which is suitable for use in a vehicle or boat."
Arkansas	"[A] battery with a core of elemental lead and a capacity of six or more volts."
	"[A]ny battery which is primarily composed of both lead and sulfuric acid, with a capacity of six volts or

California	more, and which is used for any of the following purposes: 1) as a starting battery which is designed to deliver a high burst of energy necessary to crank an engine until it starts; 2) as a motive power battery which is designed to provide the sources of power for propulsion or operation; or 3) as a stationary standby battery which is designed to be used in systems where the battery acts as a source of emergency power, serving as a backup in case of failure or interruption in the flow of power from the primary source."
Connecticut	"[A] lead-acid battery or a motor vehicle battery."
Florida	Under the recycling provisions, the term lead-acid battery is undefined. Under the battery fee provisions, the term is defined as follows: "[T]hose lead-acid batteries designed for use in motor vehicles, vessels, and aircraft, and includes such batteries when sold as a component part of a motor vehicle, vessel, or aircraft, but not when sold to recycle components."
Georgia	The term lead-acid battery is undefined, however, the law applies only to "lead-acid vehicle batteries. . ."
Hawaii	The term lead-acid battery is undefined, however, the law appears to apply to "motor vehicle or other lead-acid batteries. . ."
Idaho	"[A] battery with a core of elemental lead and a capacity of six or more volts which is suitable for use in farm equipment, construction equipment, a motor vehicle or a boat. Batteries only suitable for motor cycles, off-road recreation vehicles or lawn and garden equipment are exempt from the fees in this chapter."
Illinois	"[A] battery containing lead and sulfuric acid that has a nominal voltage of at least six volts and is intended for use in motor vehicles."
Indiana	"[A] battery that: 1) contains lead and sulfuric acid; and 2) has a nominal voltage of at least six volts."
Iowa	The term lead-acid battery is undefined.
Kansas	"Any battery that consists of lead and sulfuric acid, is used as a power source, and has a capacity of six (6) volts or more."
Kentucky	The term lead-acid battery is undefined.
Louisiana	The term lead-acid battery is undefined, however, the law applies to "motor vehicle batteries or other lead-acid batteries. . ."
Maine	"[A] device designed and used to store electrical energy through chemical reactions involving lead and acid."
Massachusetts	The term lead battery is undefined and unqualified in the regulation.
Michigan	"[A] storage battery, that is used to start an internal combustion engine or as the principal electrical power source for a vehicle, in which the electrodes are grids of lead containing lead oxides that change in composition during charging and discharging, and the electrolyte is dilute sulfuric acid."
Minnesota	The term lead-acid battery is undefined and unqualified in the law.
Mississippi	The term lead-acid battery is undefined, however, the law applies to "motor vehicle batteries or other lead-acid batteries. . ."
Missouri	"[A] battery designed to contain lead and sulfuric acid with a nominal voltage of at least six volts and of the type intended for use in motor vehicles and watercraft."
Nebraska	
Nevada	The term lead-acid battery is undefined, however, the law applies to "motor vehicle batteries."
New Hampshire	The term lead-acid battery is undefined, however, the law applies to "motor vehicle or wet cell batteries."
New Jersey	"[A] lead-acid electric storage battery designed for use in motor vehicles, aviation equipment or marine vessels."
New Mexico	The term lead-acid battery is undefined and unqualified in the regulation.

New York	"[A]ny battery with a capacity of six or more volts which contains lead and sulfuric acid and which is used as a power source in a vehicle."
North Carolina	The term lead-acid battery is undefined, however, the law applies to "motor vehicle batteries or other lead-acid batteries. . ."
North Dakota	The term lead-acid battery is undefined.
Oklahoma	"[A] lead-acid electrical device used in boats, planes and motor vehicles."
Oregon	The term lead-acid battery is undefined.
Pennsylvania	The term lead-acid battery is undefined, however, the law applies to ""motor vehicle batteries or other lead-acid batteries. . ."
Rhode Island	"[B]atteries used in any vehicle, or of a capacity of six (6) volts or more, and of one hundred fifty (150) pounds or less in weight, and like batteries in stationary uses."
South Carolina	"[A]ny battery that consists of lead and sulfuric acid, is used as a power source, and has a capacity of six volts or more, except that this term shall not include a small sealed lead-acid battery. A small sealed lead-acid battery weighs twenty-five pounds or less, and is used in non-vehicular, non-SLI (starting lighting ignition) applications."
South Dakota	The term lead-acid battery is undefined.
Tennessee	The term lead-acid battery is undefined, however, the law applies to "automobile batteries."
Texas	Under the recycling provisions, the term lead-acid battery is undefined, however, the law applies to "motor vehicle batteries or other lead-acid batteries. . ." Under the battery fee provisions, the term is defined as follows: "any battery with a capacity of six or more volts which contains lead and sulfuric acid."
	The term lead-acid battery is undefined, however, the law applies to any "motor vehicle battery or other lead-acid battery. . ."
Utah	"Lead-acid battery means a battery that consists of lead and sulfuric acid and is used as a power source."" Nevertheless, the law exempts small sealed lead-acid batteries (weighs less than twenty-five pounds, is used in non-vehicular applications and is not used as the principal source for transportation, including automobiles, motorcycles, and boats)."
Vermont	
Virginia	The term lead-acid battery is defined as "any wet cell battery."
Washington	"[B]atteries capable for use in any vehicle, having a core consisting of elemental lead, and a capacity of six or more volts."
West Virginia	"Lead-acid battery means an encasement which contains or contained lead and sulfuric acid to produce an electrical charge."
Wisconsin	The term lead-acid battery is undefined, however, the law applies to any "motor vehicle battery or other lead-acid battery. . ."
Wyoming	The term lead-acid battery is undefined, however, the law applies to any "motor vehicle battery or other lead-acid battery. . ."

Summary (Total = 42 States and 1 City)

_ 37 States and 1 City with the BCI Model (with and without deposit);

_ 7 States with a $5 deposit in lieu of trade-in requirement; and

_ 2 States with a $10 deposit in lieu of trade-in requirement.

_ 5 States with a ban on municipal solid waste disposal (landfills and/or incinerators)

To view the *State Lead-Acid Battery Laws Point-of-Sale Language* **CLICK HERE**.

Click here to find out where to take used lead-acid batteries for recycling.

Battery Council International

---

401 North Michigan Avenue, 24th Floor
Chicago, Illinois 60611-4267
Phone: 312-644-6610
Fax: 312-527-6640

contact the info@batterycouncil.org
with any comments regarding this site

XX. EVALUATION, INSPECTION AND ABATEMENT OF PAINT, DUST AND SOIL

A. General

Whether an owner or operator of a building, structure or residence has an obligation to evaluate, inspect for or abate lead-based paint, dust or soil is determined by the various federal, state and local laws described in other chapters. For privately owned industrial and commercial buildings and structures, there is no mandatory requirement, although employers must comply with OSHA's workplace exposure standards described in Chapter XI. For most privately-owned residences, there is also no mandatory requirement.

Nevertheless, a growing number of residential properties are being evaluated, inspected and abated under a variety of laws. The largest programs are those sponsored by HUD for federally owned or assisted housing. Under Title X of the Residential Hazard Reduction Act, lead-based paint in public and Indian housing must be abated when the housing is renovated or remodeled. Many evaluations, inspections and abatements are being triggered under state and local laws and regulations based on reduced action levels resulting from the Statement issued by the CDC in 1991. The EPA's disclosure rule requiring that sellers of pre-1978 residential property allow buyers to arrange a risk assessment or inspection, if not waived, may result in abatements, even though the rule does not require sellers to do so.

This Chapter briefly describes common methods to evaluate, inspect and abate paint, dust and soil, based primarily on the Guidelines for the Evaluation and Control of Lead-Based Paint Hazards in Housing issued in 1995, and partially updated in 1997, by the U.S. Department of Housing and Urban Development (the "HUD Guidelines"). The HUD Guidelines are available on HUD's website. The HUD Guidelines are not themselves enforceable by law unless a federal, state or local law or regulation requires that they be followed.

B. Definitions

Under TSCA and other federal laws, lead-based paint is defined as paint containing 1.0 mg/cm^2 as an area concentration or 5 mg/kg (i.e., 0.5% or 5,000 ppm) as a concentration by dry-weight. This standard, of course, applies to dried, cured paint on surfaces. The CPSC definition of 0.6 mg/kg (600 ppm) applies to manufacture and sale of lead-based paint as a consumer product and not to the millions of gallons of lead-based paint applied prior to the ban in 1978. State and local laws sometimes use different definitions.

Under TSCA, a lead-based paint hazard is any condition that causes exposure to lead from lead-contaminated dust, bare lead-contaminated soil, or lead-based paint that is deteriorated, or intact lead-based paint present in accessible surfaces, friction surfaces or impact surfaces that would result in adverse human health effects. Under this definition, the HUD Guidelines point out that intact lead-based paint on most walls and ceilings is not considered a hazard, although the condition of the paint should be monitored to ensure that it does not become deteriorated. Most efforts to reduce lead hazards in housing are aimed at controlling lead-based paint hazards, with the greatest concern being focused on those to which children are most frequently exposed. The hazards to children include (1) dust and bare soil that can be ingested by hand to mouth activities; (2) deteriorated surfaces; (3) chewable surfaces such as windowsills; (4) and friction and impact surfaces that may generate paint chips more readily than less accessible surfaces.

C. Standards

When performing an evaluation, inspection, or abatement, it is necessary to comply with strictest applicable standard imposed by any applicable federal, state or local authority.

As stated earlier, the federal standard for lead-based paint is 1.0 mg/cm^2 or 0.5% by weight. Some states have higher (up to 1.2 mg/cm^2) or lower (down to 0.7 mg/cm^2) standards.

The EPA has defined a "paint-lead hazard" as any of the following:

(a) Any lead-based paint on a friction surface that is subject to abrasion and where the lead dust levels on the nearest horizontal surface underneath the friction surface (e.g., the window sill, or floor) are equal to or greater than the dust-lead hazard levels identified as dust-lead hazard below.

(b) Any damaged or otherwise deteriorated lead-based paint on an impact surface that is caused by impact from a related building component such as a doorknob that knocks into a wall or a door that knocks against its doorframe.

(c) Any chewable lead-based painted surface on which there is evidence of teeth marks.

(d) Any other deteriorated lead-based paint in any residential building or child-occupied facility or on the exterior of any residential building or child-occupied facility.

A "dust lead hazard" is surface dust in a residential dwelling or child-occupied facility that contains a mass-per-area concentration of lead equal to or exceeding 40 ug/ft^2 on floors or 250 ug/ft^2 on interior windows based on wipe samples.

A "soil lead hazard" is bare soil on a residential real property or on the property of a child-occupied facility that contains total lead equal to or exceeding 400 parts per million in a play area or average of 1,200 parts per million of bare soil in the rest of the yard based on soil samples.

Some state and local standards may differ from the EPA's standards.

The federal standards for waste disposal are determined by TCLP tests as described in Chapter IX. Again, some state standards may be different.

The worker protection standards in OSHA's construction industry standard (see Chapter XI) are based on airborne lead concentrations, not concentrations in paint, dust or soil. Thus, compliance with OSHA's standards can and must be achieved regardless of the presence of lead at high or low concentrations by assuring that the lead does not become airborne as dust or fumes in excess of the permissible exposure limit (PEL). Conversely, OSHA's regulations do not directly require abatement of lead in paint, dust or soil, although abatement is an option that can be used by employers to reduce or eliminate sources of airborne lead.

See the description of OSHA's Construction Industry Standard in Chapter XI (D) for further information about the obligations of abatement contractors and other employers to protect their workers from lead exposure.

D. Risk Assessments

According to the HUD Guidelines, the term "risk assessment" means an on-site investigation of a residential dwelling for lead-based paint hazards and includes a visual inspection; limited environmental samplings of dust, soil and deteriorated paint; and a report of the results that identifies acceptable abatement and interim control strategies for controlling any lead-based paint hazards identified. A "paint inspection" is a surface-by-surface investigation of all painted surfaces, interior and exterior, in common areas of multifamily buildings as well as dwelling units, using portable X-ray fluorescence (XRF) analyzers and/or laboratory analysis of paint samples to determine the presence of lead-based paint and provision of a report on the results. Risk assessments and paint inspections can be combined.

A risk assessment must be performed by a person certified as a risk assessor. A risk assessment can be performed together with an inspection if personnel certified to perform both tasks are available. However, a risk assessment may make an inspection unnecessary if no lead-based paint hazards, as defined by applicable regulations, are found during the risk assessment. On the other hand, if obvious hazards are present, it is permissible to proceed directly to inspection and to controls which may include abatement. In that event, however, a risk assessment should be performed as part of the clearance process after the interim controls have been implemented or the abatement work performed.

Two levels of risk assessments can be performed. One alternative is a lead hazard screen for properties in good to fair condition. Building Condition Form 5.1 in HUD's Guidelines provides criteria that can be used to determine whether a property is in good enough condition to qualify for a screen risk assessment. See Appendix A at the end of this chapter. The HUD Guidelines state that screen methods should not be used in buildings in poor condition, especially structures built before 1960. Thus, they are most useful for properties built between 1960 and 1977, when use of lead-based paint in residential properties had declined to low levels, although not banned until 1978. Table 5.1 from the HUD Guidelines showing the declining use of lead-based paint prior to 1978 appears as Appendix B at the end of this Chapter.

The other alternative is a full risk assessment, which is a very detailed procedure and includes some sampling and testing of lead-based paint as well as dust or soil.

Risk assessment procedures include collection of information about the property and its occupants and uses and about the maintenance and cleaning practices used at the property. The risk assessor will perform an onsite visual assessment. The risk assessor will also collect samples of deteriorated paint, surface dust and soil. The risk assessor will then send them for testing to a laboratory accredited under the EPA's National Lead Laboratory Accreditation Program (NLLAP). Portable XRF testing can be used if the deteriorated paint chip samples have enough surface with all layers present.

When the laboratory results are received, the risk assessor will evaluate all the information obtained during the assessment and will prepare a report either concluding that no hazards were identified or identifying the hazards. Depending on the scope of the assessment, the assessor may also recommend control options from interim controls to complete abatement. If requested, the assessor may also provide cost estimates.

A typical risk assessment does not usually include air or water sampling and testing or investigation of nonobvious exposure sources. However, if a child with a blood lead level in excess of applicable standards is involved, the investigation may add air, water and other potential sources and exposure pathways. In these situations, the assessor will report the sources identified, but may not be able to recommend control measures if the source is outside the property. The HUD Guidelines recommend that potential sources be identified in the report and local health authorities be notified.

E. Inspection, Sampling and Testing Methods

Inspection and testing for lead-based paint can be accomplished by several methods including methods issued by ASTM and the National Institute of Standards and Technology (NIST). Chemical testing with sodium sulfide has been used for some years, but this method results in high levels of errors. The HUD Guidelines recommend against use of chemical test kits using sodium sulfide or sodium rhodizonate.

X-ray fluorescence (XRF) analyzers are widely used for detection of lead-based paint in housing. The HUD Guidelines recommend that XRF analyzers be used in accordance with the XRF Performance Characteristic Sheet, U.S. Nuclear Regulatory Commission, and state radiation regulations.

XRF instruments are portable. They report detection of lead measured as mg/cm^2. They do not report percent by weight. Although more reliable than chemical testing, XRF equipment must also overcome some problems to obtain accurate results. The equipment must be carefully calibrated, reading times must be monitored and adjusted, subtractions must be made for readings based on substrate rather than paint, and adjustments must be made for age of the radioactive source material.[1] K XRF equipment is considered to be more sensitive than L XRF equipment. EPA field test reports indicated standard corrections of XRF readings of 0.25 mg/cm^2 for wood or drywall substitutes, 0.21 mg/cm^2 for steel, and 0.49 for concrete, brick and similar substances.

To demonstrate that sampling has been performed properly, it is best to develop a sampling plan that will show that areas likely to contain lead were not overlooked. For example, Chapter 7 of HUD's Guidelines recommends that inspectors of multiunit properties establish testing combinations based on room equivalent, building components and substances when using XRF equipment. Recommendations are also made for reading times, calibration check readings, number of readings per testing combination, substrate corrections, and classification of XRF test results. Chapter 7 also briefly discusses paint chip collection and laboratory analysis and evaluation and documentation of the inspection.

Among the factors which complicate inspection are the painting and repainting histories of many residences. The difference between area concentration measurement and a weight percent measurement is also difficult. For example, a wall with paint containing at 1.0 mg/cm^2 or more of lead will be reported as lead-based paint, although the weight percent may be less than 0.5%.

When XRF equipment test results are marginal, the tests can be repeated or the inspector can collect paint chip samples and send them to a laboratory for testing before making a decision to undertake costly abatement work. The laboratory should be recognized under

[1] XRF analytical equipment does not present an imminent hazard when properly used. Use, storage, and disposal is subject to regulation by the Nuclear Regulatory Commission (NRC) and state agencies in states which have agreements with the NRC. Chapter 7, Section 7 of the HUD Guidelines provides information on radiation hazards.

EPA's National Lead Laboratory Accreditation Program. A laboratory can report test results either as area concentration, if samples are carefully selected and measured, or as concentration by weight. Laboratories have used two primary instruments to analyze for lead: flame atomic absorption spectrophotometric (FAAS) equipment and graphite furnace atomic absorption spectrophotometric (GFAAS) equipment. FAAS equipment can detect with reasonable reliability at about 0.001 mg. GFAAS equipment can test with reasonable reliability to levels less than one part per billion. Laboratories may also use Inductively Coupled Plasma-Atomic Emission Spectroscopy (ICP-AES), Anodic Stripping Voltammetry (ASV), and Potentiometric Stripping Voltammetry (PSV).

As the volume of lead testing has grown, use of GFAAS equipment has grown, especially at laboratories which test for drinking water as well as blood, paint, dust and soil. GFAAS equipment is manufactured by Perkin-Elmer Corporation and has also been made by Jarrell Ash (now part of Genesis Laboratory Systems, Inc.).

In a lawsuit, a county court in New York ordered abatement where K-shell XRF readings exceeded 1.0 mg/cm^2, after substrate corrections, even though laboratory atomic absorption tests showed results less than 0.5% by weight.[2]

F. Interim Controls

TSCA defines "interim controls" as a set of measures designed to reduce temporarily human exposure or likely exposure to lead-based paint hazards, including specialized cleaning, repairs, maintenance, painting, temporary containment, ongoing monitoring of lead-based paint hazards or potential hazards, and the establishment and operation of management and resident education programs. The term is used in contrast with the term "abatement" which means a set of measures to permanently eliminate lead-based paint hazards.[3]

Interim controls can be implemented in anticipation of abatement. Under Subchapter IV of TSCA, interim controls can also be combined with maintenance, monitoring

---

[2] H&H Equities v. Baez, N.Y. Civil Court, Bronx County, 1995.

[3] 15 U.S.C. § 2681(1) and (8).

and reevaluations to provide a remedy which can be effective indefinitely. See Chapter 11 of the HUD Guidelines.

The HUD Guidelines warn that interim controls should not be used in housing which has substantial structural defects or major components that are seriously deteriorated or subject to excessive moisture. Any structural problems should be repaired before interim controls are implemented. Interim controls should not be used when abatement is required by law.

The HUD Guidelines provide the following list of interim control measures:

- Repairing all rotted or defective substrates that could lead to rapid paint deterioration (repairing defective building systems that cause substrate damage may be a prerequisite for effective interim control but is outside the scope of interim control per se).
- Paint film stabilization - stabilizing all deteriorated lead-based paint surfaces by removing deteriorating paint and repainting.
- Friction and impact surface treatments - treating floors, interior window sills and window troughs so that they are smooth and cleanable.
- Treating friction and impact surfaces, such as windows, doors, stair treads, and floors, when they are generating lead-based paint chips or excessive levels of leaded dust that cannot be controlled with ordinary cleaning.
- Treating protruding, accessible surfaces, such as interior window sills, where lead-based paint may be present and there is either visual or reported evidence that children are mouthing or chewing them.
- Treating all bare soil containing excessive levels of lead.
- Dust removal and control - cleaning surfaces to reduce levels of leaded dust to acceptable levels, including cleaning carpets, if they are contaminated.
- Educating residents and maintenance workers on how to avoid lead poisoning.
- Conducting reevaluations by certified individuals, ongoing monitoring by owners, and observation by residents.

After completion of the control measures, the HUD Guidelines prescribe that an independent, certified inspector technician or risk assessor carry out a clearance examination. Property owners should implement an active maintenance regime to assure that the property continues to be free of hazards.

Interim controls and subsequent maintenance, monitoring and revaluations must be systematic and detailed. For example, the HUD Guidelines prescribe methods for paint film stabilization and provide useful information such as reasons for coating failures, substrate condition and repairs, general paint application guidelines, worksite preparation, paint removal methods, surface cleaning, primer topcoats, cleaning and clearance, and evaluation and monitoring. A quick "paintover" is unlikely to meet the standards in the HUD Guidelines which recommend even such specifications as primer thickness, topcoat thickness and total coating thickness.

To the extent applicable, owners and contractors implementing interim controls should use the methods prescribed for abatement work. For example, although interim control work should ordinarily generate less dust, debris and fumes than abatement work, steps should be taken to protect occupants and workers when interim control work may create such exposures.

Before implementing interim control measures, a review should be made to assure that the proposed measures meet state and local laws and regulations as well as the federal laws and regulations.

G. Abatement Methods

Property owners, contractors and others should keep in mind that lead has rarely presented an imminent hazard of death or serious symptomatic illness in recent decades. Today, improperly performed abatement work is perhaps the most serious risk associated with lead-based paint because untrained workers using improper methods risk daily exposure for long time periods. Lead-based paint can lawfully be managed in place, so there is no reason for abating lead-based paint by substandard methods that may risk worker health.

Recognizing the potential hazards of abatement work, federal and state laws and regulations are imposing increasingly strict requirements on abatement contractors. The EPA's regulations on accreditation of training providers and laboratories and the certification of contractors are described in Chapter XVII.

The first steps in abatement are to inspect and prepare the worksite and to arrange for occupant and worker protection. An inspection should determine, among other things, the extent of precleaning by high efficiency particulate air (HEPA) vacuum equipment that may be needed and how the site can best be controlled during the work. Plans should be made for containment of the work area by such means as plastic sheeting and posting warning signs restricting entry by unauthorized personnel. Arrangements should either be made to relocate building occupants, if required, or to assure that they do not enter the work area and are protected from dust and other possible hazards if they remain. Plans should be made to protect workers from exposure by the means described in Chapter XI including respirators and other personal protective equipment.

Abatement can be accomplished by several methods. They include the following:

- Building component replacement
- Enclosure systems other than encapsulation
- Encapsulation with durable liquid applied coatings or adhesively bonded covering materials
- Paint and dust removal, onsite or offsite
- Soil removal or covering

Removal and Replacement: In some older properties, removal and replacement of windows, doors, railings, baseboards and other building components may be quicker, safer and more economical than other methods. The objective is to accomplish the removal while generating minimal dust and debris and also to seal and arrange for lawful disposal of the components.

Enclosure: Enclosure is a relatively quick and safe, but costly, abatement method if the painted substrate is sound. Gypsum board, fiberboard, wood panels with veneers, plastic laminates, tiles and synthetic brick and stone materials are examples of durable materials with approximately 20-year design lives that are recognized for enclosure use. The material must be firmly affixed and seams must be sealed. The HUD Guidelines prescribe labelling of the surface to be enclosed with a warning: "Danger: Lead-Based Paint." and preparation of a durable floor plan drawing to warn of the locations of lead-based paint. The HUD Guidelines also recommend periodic monitoring of the enclosed surfaces.

Encapsulation: Encapsulation is a process that makes lead-based paint inaccessible by providing a barrier between lead-based paint and the environment by using a liquid applied coating, such as an epoxy, or an adhesively bonded covering material, such as a fibermat or vinyl floor tile. Encapsulation has been regarded with caution by environmental government agencies. However, encapsulant manufacturers have provided demonstrations that properly formulated materials can have satisfactory adhesion, protection and durability. Chapter 13 of the HUD Guidelines considers encapsulation to be an acceptable method if (1) the product or system is warranted by the manufacturer to perform for at least 20 years; (2) selection and use follows the manufacturer's recommendations, procedures and precautions; (3) patch testing is completed successfully; (4) the property owner or local government agency conducts surface-by-surface periodic visual monitoring; and (5) failures are repaired as soon as possible, as provided in the Guidelines. Encapsulation should not be used on friction surfaces, deteriorated components, severely deteriorated paint films, or surfaces in which there is a known incompatibility between two existing coating layers. Fire, health and environmental requirements must be met when working with chemical encapsulant liquids. When successfully used, encapsulatants offer the advantages of low cost and relative speed and ease of application.

Paint Removal: Federal and state government agencies warn that paint removal may not be the best and, in some circumstances, even a suitable abatement method. HUD, OSHA and other government agencies have prohibited paint removal methods such as the following:

- Open flame burning or torching
- Machine sanding or grinding without HEPA exhaust ventilation

- Abrasive blasting or sandblasting without HEPA vacuum exhaust equipment and other appropriate containment under negative pressure
- Hydroblasting and high-pressure water washing without collection and containment of all water effluent.

HUD's Guidelines and the rules of other government agencies also discourage or restrict dry scraping which generates paint chips and dust. Wet scraping is recommended except near electrical equipment and when using a heat gun. HUD's Guidelines discourage and some state agencies prohibit the use of chemical paint removers containing methylene chloride, a suspect carcinogen. OSHA requires air-supplied respirators or self-contained breathing apparatuses if they are used.

HUD's Guidelines recommend the following paint removal methods:

- Electric powered flameless heat guns when used with OSHA required fire protection equipment and air-purifying respirators with a HEPA-filtered cartridge and an organic vapor cartridge
- Disc, orbital or vibrating sander equipment with shrouds or containment systems under a partial vacuum which collects and exhausts air through a HEPA filter
- Wet scraping, except near electrical equipment
- Abrasive blasting with a shroud under a vacuum attached to the blast head so that all exhaust air is passed through a HEPA filter
- Vacuum needle gun equipment with shroud and air exhaust through a HEPA filter
- Chemical removal, especially in historical sites, if done with adequate skill and use of respirators, gloves and other personal protective equipment to guard against splashes of caustic materials and exposure to vapors of hazardous materials such as methylene chloride. Offsite chemical removal, when feasible, is preferred

over onsite chemical removal because risks can be better controlled.

Soil Abatement: The EPA's soil abatement standards are described in Chapter XVII. Abatement methods include soil removal and replacement, soil cultivation (rototilling), soil treatment and replacement, soil cleaning, and paving with concrete or asphalt. HUD recommends that other options be considered before choosing to remove contaminated soil because of the difficulties and costs. The HUD Guidelines limit soil cultivation as an abatement strategy to soil below 1,500 ug/g, lead concentration. The Guidelines state that HUD is investigating three soil cleaning methods for possible use on residential sites: magnetic separation, froth flotation and washing. One practical alternative offered by HUD is paving with high quality concrete or asphalt. However, HUD warns that hard surfaces are not appropriate in play areas where falls are possible.

H. Cleaning

Because abatement work tends to generate dust, debris and sometimes other contamination, cleaning during abatement work and at its conclusion is essential. Otherwise, the abatement work can create conditions more hazardous than those existing prior to the abatement.

Cleaning during lead abatement work is not an easy task because the standards set by HUD and other government authorities are strict. HUD's Guidelines report that significant failure rates were encountered in clearance tests during a 1991 study by HUD. The Guidelines explain that workers are used to cleaning dust they can see, not invisible particles that are also important to remove.

Cleaning methods begin with effective site containment using plastic sheet and vacuuming with special vacuum equipment with a HEPA exhaust filter. The next steps include wet washing with trisodium phosphate (TSP) detergent solution (or another detergent designed for lead abatement) and final HEPA vacuuming. (NOTE: HEPA filters remove particles of 0.3 micron or greater from air with 99.97% efficiency.) Special attention is needed for difficult surfaces such as carpets. Debris should be collected and sealed in double 4-inch or single 6-mil polyethylene (or equivalent) plastic bags and stored for disposal. Before clearance, treated surfaces and floors should be coated and sealed. It is essential to clean the site daily during the progress of the abatement work and not wait until its completion.

The Guidelines say that some vacuums have an ultra-low penetration air (ULPA) filter that can filter out particles of 0.13 microns or greater at 99.9995% efficiency.

I. Waste Disposal

As described in Chapter XI, generation, treatment, storage, transportation and disposal of waste, both hazardous and nonhazardous, is comprehensively regulated by the SDWA, as amended by RCRA, and by corresponding state laws and regulations. Compliance with waste disposal laws and regulations is sometimes a difficult task during work to control or abate lead-based paint.

Waste generated during interim control and abatement work must be evaluated to determine whether it is among the wastes in the lists of hazardous wastes prepared by the EPA.[4] The waste must also be evaluated and, if necessary, tested to determine whether it has a characteristic of ignitability, corrosivity, reactivity or toxicity as defined in the EPA's regulations.[5] If the waste is listed or has one or more of the defined characteristics, it must be managed as hazardous waste unless an exemption is available.

It is unlikely that waste generated during interim control or abatement proceedings will be a listed waste or a reactive waste. However, some of the waste may be ignitable, corrosive or toxic. For example, some solvent wastes may be ignitable. Some caustic or acid wastes may be corrosive. Lead containing wastes from which lead leaches readily may be toxic. The tests to measure these characteristics are prescribed by the EPA's RCRA regulations.[6]

Property owners and contractors should not assume that lead containing paint or other wastes are toxic under RCRA based on XRF tests or FAAS or GFAAS laboratory tests at a

---

[4] 40 CFR § 261.30.

[5] 40 CFR § 261.30.

[6] 40 CFR Part 261.

laboratory to identify lead-based paint. The XRF tests are designed to penetrate layers of paint and the laboratory tests use nitric acid to measure total concentration. The RCRA toxic characteristic leaching procedure (TCLP) test uses a mild acid solution to measure the tendency of a hazardous substance to leach out of a waste under conditions that might exist, for example, in a landfill. The amount of lead that will leach from waste in a TCLP test is usually much lower than the total content measurable by destructive tests.

The HUD Guidelines point out that interim control waste may be eligible for the household waste exclusion under the EPA's RCRA regulations. Small quantities of waste water, less than 10 gallons, from cleaning or mopping operations can be poured down a toilet if local authorities approve of such a practice, but the local waste water treatment facility should be contacted before discharging a larger quantity.

The HUD Guidelines divide abatement waste into four categories: low lead waste, architectural components, concentrated lead waste, and other waste. The Guidelines provide practical information on management practices to be used for each of these four categories in order to comply with the EPA's RCRA regulations as well as some precautions about the need to comply with state and local laws and regulations.

The EPA's hazardous waste generator regulations at 40 CFR, Part 262 contain an exemption that can facilitate compliance at small sites. If less than 100 kg (220 pounds) of hazardous waste is generated per month, the waste is "conditionally exempt" and can be sent to a state permitted nonhazardous waste landfill without use of a hazardous waste manifest. Up to 1,000 kg (2,200 pounds) can be stored onsite at any one time. The EPA's regulations also contain an exemption for generators producing more than 100 kg per month, but less than 1,000 kg per month. However, generators in that category must comply with most of the requirements applicable to generators of larger quantities, so it offers few advantages. Generators of more than 100 kg per month of hazardous waste must comply with several requirements:

1. Obtain a generator number
2. Train workers about RCRA compliance

3. Store hazardous waste in tanks or containers labelled "Hazardous Waste" and monitor the storage accumulation times. If more than 1,000 kg (one ton) of hazardous waste is generated at the site, do not store longer than 90 days. (Generators who wish to store longer than 90 days must obtain a storage permit and comply with extensive requirements.)
4. Hire a reputable waste transporter who has an EPA permit for the hazardous waste and can dispose of the waste at reasonable cost at a permitted disposal facility. Be sure about the reputations, permits and costs of both the transporter and the disposal facility because the generator is likely to be held responsible if either mishandles the waste.
5. Properly package, mark, label or placard, and complete proper shipping papers for the hazardous waste. The Uniform Hazardous Waste Manifest Form 8700-22 (and, if necessary, continuation sheet form 8700-22A) must be used unless the conditional exemption for less than 100 kg per month of hazardous waste generated at the site applies. After signing the Manifest, the generator must obtain the transporter's signature, and the transporters must arrange for the disposal facility to sign and return the manifest within 35 days.
6. If disposal is at a landfill, comply with the EPA's land disposal notification, certification and other requirements.
7. Maintain all records for at least three years and comply with EPA periodic reporting requirements if more than 1,000 kg per month of hazardous waste is generated at the site.

Chapter IX contains a description of lead-based paint waste disposal information updated since the HUD Guidelines were published. Of specially practical importance are the amendments adopted by the EPA in 2001 facilitating disposal of residential lead-based paint waste at construction and demolition (C&D) landfills.

J. Clearance

When a contractor reports that abatement work is done, including cleaning, the next step is clearance. If the abatement work is performed for a government agency such as HUD, or under order of a state health department, clearance must be performed in accordance with the agency requirements. Government agencies customarily require a clearance inspection by their own personnel or by a certified risk assessor or inspector. When abatement is arranged privately, the property owner or operator can choose whether to rely on the abatement contractor, conduct a self-evaluation, or hire a certified risk assessor or inspector. Factors in the decision include the extent and difficulty of the work, the reputation of the abatement contractor, and the relative cost of available risk assessors or inspectors.

The HUD Guidelines state that clearance dust sampling should be performed no sooner than one hour after completion of the final cleanup in order to allow dust to settle. Clearance sampling and testing is for settled dust, not airborne dust, since the main source of lead exposure for children is ingestion of lead dust. Entry into the containment area should be prohibited during the one hour waiting period. NOTE: The Guidelines explain that air sampling is used in clearance procedures for asbestos abatement because asbestos fibers take a long time to settle out of air. However, air sampling is not a useful clearance tool for lead hazard control work because lead particles settle much faster than asbestos fibers.

The clearance reviewer will begin with a visual inspection of the work to assure that it was properly performed. If the abatement work was paint removal, the inspection should be done before repainting, and the reviewer should determine whether any surfaces were overlooked or paint residues remain on any surfaces. If the work was enclosure, the fasteners and sealing should be checked. If the work was encapsulation, the performance of patch tests and adhesion and sealing should be checked. The reviewer will then inspect for dust and debris, visually and by wipe sampling. The reviewer will sample locations such as window components, floors and carpets using sampling protocols. The reviewer will also obtain samples of any bare soil near the work and arrange for testing.

The HUD Guidelines provide that composite dust samples should be sent for laboratory testing by FAAS, GFAAS or ICP equipment. Homeowners and owners of small residential properties rented to tenants will find these tests to be expensive.

When samples are tested and the results are received, the reviewer will compare them to the clearance standards described earlier in this Chapter. The reviewer will then report his or her findings and recommendations to the property owner or operator. If the work fails to meet the clearance standards, it should be corrected until the standards are met. Owners or operators should be careful to retain part of the price for the work to assure that the contractor will perform all necessary corrective work.

Property owners and operators should be aware that government protocols for sampling and testing are very detailed. If the work must meet government standards, the protocols must be followed or the clearance will not be recognized. For example, minimum numbers of samples must be obtained from the surfaces in each room unit as shown in Table 15.1 from the HUD Guidelines that appears at the end of this Chapter. The proper materials must be used to obtain dust samples, and they must be placed in the proper containers to be sent to the laboratory. Composite samples can be taken and prepared, but only by proper procedures. Chain of custody procedures must be followed when samples are forwarded to the laboratory and they should be accompanied by samples spiked with known quantities of contamination as a means of verifying the accuracy of the laboratory work.

K. Renovation and Remodeling

Until 2006, the EPA had limited regulations applicable renovation and remodelling of housing that is not for the purpose of abatement, except to require renovation contractors to deliver the EPA pamphlet "Protect Your Family From Lead in Your Home." However, HUD and the EPA conducted warning campaigns which included descriptions of hazards that could result if renovation and remodeling is not performed by methods similar to those they require for abatement activities. For example, Chapter 4 of HUD's Guidelines describes studies associating renovation activity with elevated blood lead levels in children. HUD's Guidelines also contain a Table 4.2 which provides a summary of measures that HUD recommends for several types of renovation work: demolition, repainting, floor sanding, plaster repairs, window replacement and carpet removal.

See Chapter XVII for a description of the Renovation, Repair and Painting Program Amendments proposed by the EPA in 2006. 71 Fed. Reg. 1588 (Jan. 10, 2006).

The EPA announced in the Federal Register, Vol. 72, page 12582, March 16, 2007 the availability of a new study by the EPA titled "Characterization of Dust Lead Levels after Renovation, Repair and Painting Activities" and a new study by the National Association of Home Builders titled "Lead-Safe Work Practices Survey Project."

L. Forms

The HUD Guidelines contain many helpful tables and useful forms. For example, Addendum to the Guidelines contains Data Collection Forms.

M. XRF Performance Characteristic Sheets

Addendum 3 to the HUD Guidelines contains performance characteristic sheets for XRF analyzer equipment of TN Technology, Inc., Scitec Corp., Warrington, Inc., Princeton Gramma Tech, Inc., Niton Corp., Radiation Monitoring Devices, Inc. and Advanced Detectors, Inc. NOTE: See also "A Field Test of Lead-Based Paint Testing Technologies," Technical Report, EPA 747-R-95-002b, May, 1995.

APPENDIX A

Table 5.1 Percentage of All Paint That Is Lead-Based, by Year and Component Type

Component Category	Interior	Exterior
Walls/Ceiling/Floor		
1960–1979	5	28
1940–1959	15	45
Before 1940	11	80
Metal Components[1]		
1960–1979	2	4
1940–1959	6	8
Before 1940	3	13
Nonmetal Components[2]		
1960–1979	4	15
1940–1959	9	39
Before 1940	47	78
Shelves/Others[3]		
1960–1979	0	—
1940–1959	7	—
Before 1940	68	—
Porches/Others[4]		
1960–1979	—	2
1940–1959	—	19
Before 1940	—	13

[1] Includes metal trim, window sills, molding, air/heat vents, radiators, soffit and fascia, columns, and railings.

[2] Includes nonmetal trim, window sills, molding, doors, air/heat vents, soffit and fascia, columns, and railings.

[3] Includes shelves, cabinets, fireplace, and closets of both metal and nonmetal.

[4] Includes porches, balconies, and stairs of both metal and nonmetal.

Source: HUD 1990b. These data are from a limited national survey and may not reflect the presence of lead in paint in a given dwelling or jurisdiction.

Form 5.1
Building Condition Form

Condition	Yes	No
Roof missing parts of surfaces (tiles, boards, shakes, etc.)		
Roof has holes or large cracks		
Gutters or downspouts broken		
Chimney masonry cracked, bricks loose or missing, obviously out of plumb		
Exterior or interior walls have obvious large cracks or holes, requiring more than routine pointing (if masonry) or painting		
Exterior siding has missing boards or shingles		
Water stains on interior walls or ceilings		
Plaster walls or ceilings deteriorated		
Two or more windows or doors broken, missing, or boarded up		
Porch or steps have major elements broken, missing, or boarded up		
Foundation has major cracks, missing material, structure leans, or visibly unsound		
* Total number		

* If the "Yes" column has two or more checks, the dwelling is usually considered to be in poor condition for the purposes of a risk assessment. However, specific conditions and extenuating circumstances should be considered before determining the final condition of the dwelling and the appropriateness of a lead hazard screen.

XXI. STANDARDS ORGANIZATIONS

In addition to the federal and state regulatory programs described in other Chapters, several standards organizations have significant roles which effect the regulatory process. They include the American Conference of Government and Industrial Hygienists (ACGIH), ASTM International, Inc. (ASTM), the Food Codex Committee of the National Academy of Sciences (NAS), NSF International (NSF) and the American National Standards Institute (ANSI). These organizations are not government agencies, but government agencies frequently rely, in part, on their standards because of their recognized expertise.

A. American Conference of Government Industrial Hygienists

ACGIH is a national organization headquartered at 1330 Kemper Meadow Drive, Cincinnati, Ohio 45240. Tel. 513-742-2020. Fax. 513-742-3355.

For years before modern government regulatory programs were adopted, ACGIH set threshold limit values (TLVs) for workplace exposures. The TLVs set by ACGIH have been well respected and widely observed by responsible companies. As federal and state government agencies have established their major environmental and safety programs, they have continued to turn frequently to the work of ACGIH for experience and expertise.

ACGIH has established a TLV for workplace air of 0.05 mg/m^3 (TWA) that is the same as OSHA's PEL. In 1994, the ACGIH proposed a voluntary limit for blood lead content of 30 mg/100 mL of whole blood. This limit corresponded to OSHA's action level in its standards for general industry and the construction industry, but was below the personnel exposure limit (PEL) set in OSHA's standards.

B. ASTM International, Inc.

ASTM is an international standards organization. Its address is 100 Barr Harbor Drive, P.O. Box 700, West Conshohocken, PA 19428. Tel. 610-832-9585. Fax. 610-832-9555.

In 1991, HUD requested that ASTM organize a new subcommittee to develop consensus standards on the testing and abatement of lead-based paint in housing. A new Subcommittee E06.23 on Abatement and Mitigation of Lead Hazards in Buildings was formed as part of the E-6 Committee on Performance of Buildings. The Subcommittee, which included over 100 members, coordinated its work with several ASTM committees and government agencies including HUD, the CDC, the EPA, NIOSH, the National Institute of Standards and Technology (NIST), the CPSC, the Department of Defense and OSHA.

Based on the work of the E06.23 Subcommittee, ASTM has adopted numerous standards. Standards listed in the Committee website include the following:

E1605-04	Standard Terminology Relating to Lead in Buildings See also WK9418 proposed revision
E1613-04	Standard Test Method for Determination of Lead by Inductively Coupled Plasma Atomic Emission Spectrometry (ICP-AES), Flame Atom Absorption Spectrometry (FAAS), or Graphite Furnace Atomic Absorption Sepctrometry (GFAAS) Techniques
E1644-04	Standard Practice for Hot Plate Digestion of Dust Wipe Samples for the Determination of Lead
E1645-01	Standard Practice for Preparation of Dried Paint Samples by Hotplate or Microwave Digestion for Subsequent Lead Analysis
E1727-05	Standard Practice for Field Collection of Soil Samples for Subsequent Lead Determination
E1728-03	Standard Practice for Collection of Settled Dust Samples Using Wipe Sampling Methods for Subsequent Lead Determination
E1729-05	Standard Practice for Field Collection of Dried Paint Samples for Subsequent Lead Determination
E1741-00	Standard Practice for Preparation of Airborne Particulate Lead Samples Collected During Abatement and Construction Activities for Subsequent Analysis by Atomic Spectrometry
E1753-04	Standard Practice for Use of Qualitative Chemical Spot Test Kit for Detection of Lead Dry Paint Films
E1775-01	Standard Guide for Evaluating Performance of On-Site Extraction and Field-Portable Electrochemical or Spectrophotometric Analysis for Lead
E1792-03	Standard Specification for Wipe Sampling Materials for Lead in Surface Dust

E1795-04	Standard Specification for Non-Reinforced Liquid Coating Encapsulation Products for Leaded Paint in Buildings
E1796-03	Standard Guide for Selection and Use of Liquid Coating Encapsulation Products for Leaded Paint in Buildings
E1797-04	Standard Specification for Reinforced Liquid Coating Encapsulation Products for Leaded Paint in Buildings
E1828-01	Standard Practice for Evaluating the Performance Characteristics of Qualitative Chemical Spot Test its for Lead in Paint
E1864-02	Standard Practice for Evaluating Quality Systems of Organizations Conducting Facility and Hazard Assessments for Lead in Paint, Dust, Airborne Particulate, and Soil in and around Buildings and Related Structures See also WK4422 proposed revision
E1908-03	Standard Guide for Sample Selection of Debris Waste from a Building Renovation or Lead Abatement Project for Toxicity Characteristic Leaching Procedures (TCLP) Testing for Leachable Lead (Pb)
E1979-04	Standard Practice for Ultrasonic Extraction of Paint, Dust, Soil, and Air Samples for Subsequent Determination of Lead
E2051-01	Standard Practice for the Determination of Lead in Paint, Settled Dust, Soil and Air Particulate by Field-Portable Electroanalysis
E2052-99	Standard Guide for Evaluation, Management, and Control of Lead Hazards in Facilities See also WK8214 proposed revision
E2115-06	Standard Guide for Conducting Lead Hazard Assessments of Dwellings and of Other Child-Occupied Facilities
E2119-00	Standard Practice for Quality Systems for Conducting In Situ Measurements of Lead Content in Paint or Other Coatings Using Field-Portable X-Ray Fluorescence (XRF) Devices See also WK9631 proposed revision
E2120-00	Standard Practice for the Performance Evaluation of the Portable X-Ray Fluorescence Spectrometer for the Measurement of Lead in Paint Films

See also WK9632 proposed revision

E2239-04 Standard Practice for Record Keeping and Record Preservation for Lead Hazard Activities

E225204 Standard Practice for Selection of Lead Hazard Reduction Methods for Identified Risks in Residential Housing or Child Occupied Facilities

E2255-04 Standard Practice for Conducting Visual Assessments for Lead Hazards in Buildings

E2271-05a Standard Practice for Clearance Examinations Following Lead Hazard Reduction Activities in Single-Family Dwellings and Child Occupied Facilities.

An interesting article on the work of ASTM Subcommittee E 06.23, including additional standards being developed was published by ASTM in 1993. "Lead Abatement in Buildings and Related Structures," Ashley and McKnight, ASTM Standardization News, December, 1993.

C. National Academy of Sciences, Food Codex Committee

The National Academy of Sciences (NAS) is a membership organization of leading scientists headquartered in Washington, D.C. After the FDA began to regulate food additives in 1958, it requested NAS's Institute of Medicine to organize the Food Codex Committee which established the Food Chemicals Codex in 1961. The Codex has been updated several times. The Fifth Edition was published in 2003. The address is Food Chemicals Codex, Institute of Medicine, 500 Fifth Street, N.W., Washington, D.C. 2001.

The Codex contains specifications for food ingredients (including GRAS ingredients) and for color additives and food additives. In 1993, the FDA announced that the Food Codex Committee was revising its policy to set lower limits for lead and heavy metals. The Committee described the new policy as follows:

> The Committee on Food Chemicals Codex (FCC) recognizes the desirability of lowering lead exposure, especially in the case of infants and children. Overall exposure to lead is a public health

concern. Whereas diet is not the largest source of lead exposure, it is a significant one. While ingestion of FCC substances does not represent the major source of dietary lead, it is desirable to lower the lead limits for all FCC substances, particularly for those substances consumed in high amounts. Therefore, the Committee's policy is to reduce lead limits (as well as the heavy metal limit because of their interrelated nature) to the lowest extent feasible for FCC substances, especially given that more recent evidence shows deleterious neurobehavioral effects occurring in children previously considered acceptable.

In setting heavy metals and lead limits, the Committee considers the amount of a food chemical consumed, the feasibility of manufacturing a product within these limits, and the availability of analytical methods to ensure compliance. The constraints of good manufacturing practice and the availability of reliable analytical methods are often limiting factors in setting lower lead and heavy metals limits.

The Committee regards as one of its goals the assurance of the safety of properly used food chemicals. This means that FCC specifications will respond to advances in knowledge about new manufacturing methods, analytical techniques, or toxicology and safety issues.

Under the new policy, the Committee reduced the lead content specifications for a number of food ingredients as described in Chapter XIV. Accordingly, as described in Chapter XIV, the FDA reduced its specifications for color additives, food additives and GRAS ingredients.

D. NSF International

NSF is an independent, not-for-profit, testing and standards development organization headquartered at 789 N. Dixboro Road, Ann Arbor, Michigan 48113. Tel. 800-NSF-MARK. Fax. 734-769-0109. In 1985, the EPA entered into an agreement with NSF to develop health effects standards for drinking water system components. NSF adopted its Standard 61 in 1988.

In 1994, NSF added Section 9 to Standard 61 to cover faucets, drinking fountains and other endpoint plumbing devices. Section 9 sets a standard of 11 ppb or less of lead that may leach from faucets, drinking fountains and other endpoint devices. This standard is below the "action level" of 15 ppb set by the EPA under the SDWA.

NSF's website currently states that Standard 61, Section 9 has been adopted by 43 states and all major model building codes. NSF certifies faucets, drinking fountains and other products for compliance with Standard 61, Section 9.

E. American National Standards Institute

ANSI is a leading national standards organization with offices at 1819 L Street, NW, Washington, DC 20036. Tel. 202-293-8020. Fax. 202-298-9287 and 25 West 43rd Street, New York, NY 10036. Tel. 212-842-4900. Fax. 212-398-0023. As of September 12, 1995, ANSI approved NSF Standard 61, Section 9. In a related announcement, NSF said that it believed that ANSI's approval would likely result in wider acceptance and recognition by state and local plumbing codes. Those objectives have been successfully achieved.

XXII. LEAD-BASED PAINT LITIGATION

A. General

As described in earlier chapters, reports of epidemiological and other studies have been published during the last two decades indicating an association between exposure of very young children to lead-based paint hazards and neurological damage to the children's health, as manifested by lower performance on intelligence tests and behavioral and other problems. These reports led to laws and regulations relating to lead-based paint and to commencement of numerous personal injury lawsuits on behalf of children allegedly exposed to lead-based paint.

Class action lawsuits were commenced against former manufacturers of pigments contained in lead-based paint. Lawsuits were also commenced against property owners and operators seeking to hold them responsible for deterioration of lead-based paint resulting in the exposure of children. In turn, the former manufacturers and the property owners and operators asserted claims against their insurers seeking to recover defense costs and liability payments under the bodily injury and property damage provisions of general liability policies.

As growing numbers of property owners (or operators) were required to inspect for and abate lead-based paint, they sought remedies to shift the cost burden to others including their insurers, former property owners and former manufacturers of lead pigments and paints.

B. Lawsuits Against Lead Pigment Manufacturers

Plaintiff attorneys sometimes compare lead-based paint personal injury litigation to asbestos litigation which involves hundreds of thousands of lawsuits across the nation. Decisions made by the courts in asbestos lawsuits are often cited in arguing lead-based paint lawsuits as well as decisions from the diethylstilbestrol (DES), intrauterine device (IUD) and other mass tort cases.

There are, however, several important differences between other mass tort litigation and lead-based paint litigation. Lead-based paint manufacturers were very numerous and their products were sold over many years for industrial and commercial as well as retail users. Most lead-based paint for residential use was sold at retail in small quantities to small paint contractors, home owners and tenants. Several layers of lead-based paints were often applied to the same surfaces. Many years have passed since lead-based paints declined in use for residential purposes and then were banned in 1978 from sale as consumer products. Thus, plaintiffs seldom have records or other information sufficient to identify sellers and manufacturers of lead-based paint.

Plaintiff attorneys attempted to overcome this evidential problem by commencing class action lawsuits against several large companies which allegedly were producers of most of the lead pigments used in paints from 1900 to 1977 and their trade association, The Lead Industries Association, Inc. (LIA). Based on theories successfully used in the DES cases, the plaintiff attorneys sought to eliminate the manufacturer identification requirement by alleging grounds for market share, alternative or enterprise (concerted wrongful action) liability. However, their arguments were rejected by federal courts of appeals.[1]

One state appellate court allowed the City of New York to proceed based on allegations of concerted action to conceal hazards and mislead the public and the government.[2] By 1995, all claims in this lawsuit had been dismissed except fraud claims which the trial court refused to dismiss, although the City admitted that it had not purchased the lead-based paint in the buildings it owns by foreclosure.

Plaintiff attorneys also commenced class actions in state courts based on theories of market share, alternative and enterprise liability, and met with little early success.[3] A success in Ohio was limited by adoption of product liability legislation.

In the new century, plaintiff attorneys' efforts began to achieve some success. The Wisconsin Supreme Court held that liability could be imposed on lead paint and pigment manufacturers based on a risk contribution theory that is similar to market share liability.[4]

---

[1] City of Philadelphia v. Lead Industries Association, Inc., 994 F.2d 112 (3d Cir. 1993); Santiago v. Sherwin-Williams Company, 3 F.3d 546 (1st Cir. 1993); Jefferson v. Lead Industries Association, 106 F.3d 1245 (5th Cir. 1997).

[2] City of New York v. Lead Industries Association, Inc., 190 A.D.2d 173, 597 N.Y.S. 2d 698 (1st Dept. 1993).

[3] Jackson v. Glidden Company, 98 Ohio App. 3d 100, 647 N.E. 2d 879 (1994), 2007 Ohio App. LEXIS 268 Jan. 25, 2007; Skipworth v. Lead Industries Association, 547 Pa. 224, 690 A.2d 169 (1997); Brenner v. Lead Industries Association, 263 A.D. 2d 165; 699 N.Y.S.2d 848 (N.Y. App. Div. 1999).

[4] Thomas v. Mallett, 285 Wisc. 2d 236, 701 N.W.2d 523 (2005); City of Milwaukee v. NL Industries, 278 Wisc. 2d 313, 691 N.W.2d 888 (2004); review dismissed 703 N.W.2d 380 (2005).

However, they encountered losses in other courts.[5]

Recognizing that they could benefit from government sponsorship, plaintiff attorneys agreed to work for the Attorney General of the State of Rhode Island in a lawsuit against former lead pigment and paint manufacturers seeking to recover the cost of removal of lead-based paint in buildings throughout the State as well as other damages. The lawsuit alleged, among other things, that the defendant manufacturers should be held liable on a public nuisance theory although the traditional requisites did not initially appear to be present. The Superior Court in Rhode Island, however, held that the Attorney General had adequately asserted a cause of action for public nuisance and rendered several subsequent decisions including a decision excluding evidence which the defendant pigment manufacturers sought to submit.[6] The jury at a first trial could not agree on a verdict. The jury in a second trial was reportedly also unable to agree during early deliberations, but eventually returned a verdict finding the defendant manufacturers liable for having sold a product that became a public nuisance. The jury's verdict was mentioned in a subsequent decision involving insurance coverage for public nuisance claims.[7] The Superior Court denied retrial motions in February, 2007.

Lawsuits based on a public nuisance theory have also been filed in other states. In California, a lawsuit in which summary judgment had been granted to defendant manufacturers was reversed by a Court of Appeals and allowed to proceed on a public nuisance theory and several other causes of action.[8] An intermediate appellate court in New Jersey allowed municipalities to proceed on a public nuisance theory. In re Lead Paint Litigation, A-1946-02T3,

---

[5] Brenner v. Lead Industries Association, 288 A.D.2d 869, 732 N.Y.S.2d 799 (N.Y. App. Div. 2001); City of Chicago v. American Cyanamid Company et al., 355 Ill. App. 3d 209, 833 N.E.2d 126, 2005 Ill. App. LEXIS 14 (2005); City of St. Louis v. Lead Industries Association et al., No. 002-245 (Mo. Cir. Ct., Jan. 18, 2006). In January 2007, an appellate court in Ohio again denied enterprise, market share and alternative liability claims in Jackson v. Glidden, Inc., 2007 Ohio App. LEXIS 268, Jan. 25, 2007.

[6] Rhode Island v. Whitehouse, 2001 R.I. Super. LEXIS 37 (April 2, 2001); Rhode Island v. Lead Industries Association, 2004 R.I. Super. LEXIS 191 (Nov. 9, 2004).

[7] Certain Underwriters at Lloyds v. Millenium Holdings LLC, 13 Misc.3d 1204 A, 2006 N.Y. Misc. LEXIS 2326 (Aug. 8, 2006).

[8] County of Santa Clara v. Atlantic Richfield Company et al., 137 Cal. App. 4th 292, 40 Cal. Rep. 313, 2006 Cal App. LEXIS 293 (March 3, 2006). The California Supreme Court denied certioriari on June 21, 2006.

2005). In Wisconsin, an intermediate appellate court reversed a dismissal of public nuisance, conspiracy and restitution claims, and the Wisconsin Supreme Court declined review.[9]

C. Lawsuits Against Property Owners and Operators

Plaintiff attorneys have commenced numerous lawsuits against property owners and operators, primarily alleging claims on behalf of children of tenants. The lawsuits seek to impose liability for negligence, strict liability, liability based on failures to perform lease obligations, and liability based on other theories such as nuisance and fraud.

Negligence. The theory of negligence is often used in lawsuits against landlords. The plaintiff attorney must prove duty, breach and causation and must withstand the defense of contributory negligence. Since the 1960s, some courts have held landlords liable for injuries to allegedly lead-poisoned children based on negligent failure to keep leased premises in proper repair and that notice of deteriorating paint in pre-1978 housing is sufficient to invoke a landlord's repair obligation.[10] However, some courts have held that notice of deteriorated paint is insufficient to hold a landlord responsible for failure to abate lead-based paint.[11] A conclusory assertion that lead-based paint in older homes is commonly known is also insufficient.[12] These decisions were not surprising since there has been no residential use of lead-based paint for several decades and even middle-aged landlords have never seen a can of lead-based paint. As discussed later in this chapter, the liability of landlords may be affected by a statute, ordinance or code requiring inspection for, or abatement of, lead-based paint.

---

[9] City of Milwaukee v NL Industries, Inc. et al., 278 Wisc. 2d 313, 691 N.W. 2d 888, 2004 Wisc. App. LEXIS 885 (2004); review dismissed 703 N.W. 2d 380, 2005 Wisc. LEXIS 731 (2005).

[10] Acosta v. Irdank Realty, 38 Misc. 2d 859 (Sup. Ct., N.Y. Cty., 1963); Davis v. Royal-Globe, 223 So. 2d 912 (La. 1969); Juarez v. Wavecrest Management, 88 N.Y.2d 628, 649 N.Y.S.2d 115 (1996); Antwaun v. Heritage Mut. Ins, 228 Wis.2d 44, 596 N.W.2d 456 (1999).

[11] Selvy v. Beigsel, 283 Ill. App. 3d 532, 670 N.E.2d 784 (1996); Garcia Jimenez, 184 Ill. App. 3d 107, 539 N.E. 2d 1360 (1989); Winston Properties v. Sanders, 57 Ohio App. 3d 28, 565 N.E. 2d 1280 (1989); Underwood v. Risman, 605 N.E.2d 832 (Mass. 1993); Chapman v. Silber, 97 N.Y.2d 9,734 N.Y.S.2d 541 (2001); Hines v. Rap Realty, N.Y.S.2d 594 (2d Dept 1999), appeal denied 695 N.Y.S.2d 540 (1999); Andrade v. Wong, 6785 N.Y.S.2d 112 (2d Dept. 1998); Hayes v. Hambruch, 841 F. Supp. 706 (D. Md. 1994).

[12] Davis v. Royal-Globe, supra; Alharb v. Sayegh, 199 A.D. 2d 229 (3d Dept., N.Y. 1984).

In a recent case, an appellate court in New York dismissed a lawsuit on behalf of a plaintiff child against one of three defendant landlords because the plaintiff could not show that the landlord knew there was peeling paint in the apartment where the child lived for five months.[13] Another appellate court in New York held that claims could not be dismissed against landlord defendants due to fact issues whether they had actual or constructive knowledge of the lead paint condition in the apartment where the plaintiff children lived.[14]

In Ohio, a court held that sellers of a residence who had no actual knowledge of lead paint hazards and complied with the EPA's lead-based paint disclosure regulations were not required by the Residential Lead-Based Paint Hazard Reduction Act or common law to inspect for lead-based paint before sale.[15]

In a lawsuit based on negligence, a defendant landlord may not be allowed to base contributory negligence defense on negligent parental supervision because the court will not impute parental negligence to the child. However, negligence or other wrongful conduct by parents may be used to show that a landlord did not breach a duty to a child or that a landlord's failure was not the cause, or the sole cause, of a child's injuries.[16]

In New York, claims for contributions from negligent parents were not allowed.[17] However, parental contributions may be obtained in Massachusetts.[18]

Strict Liability. Failure to remove lead-based paint as required by state statutes or local ordinances has been held to be a basis for strict liability of a property owner in Massachusetts and Connecticut without proof of negligence or proof that the property owner

---

[13] Robinson v. Scafidi, 803 N.Y.S. 2d 789 (3rd Dept., 2005).

[14] Jackson v. Brown, 2006 N.Y. App. Div. LEXIS 1381 (4th Dept. Feb. 3, 2006).

[15] Munez v. J.L. Sims Co., 2003 WL 2143328 (Ohio Ct. App., June 27, 2003).

[16] Suarez v. Pyle, 823 N.Y.S.2d 728 (4th Dept. 2006); Davis v. Royal-Globe, supra; Alharb v. Sayegh, supra; Brown v. Marathon Realty, Inc., supra; Hill v. City of New York, 201 A.D. 2d 329 (1st Dept., N.Y., 1994).

[17] Franklin v. Krumanocker, 114 A.D.2d 611 (N.Y.3d Dept. 1985).

[18] Ankiewicz v. Kinder, 408 Mass. 792 (1990).

knew there was paint containing dangerous levels of lead on the premises.[19] In Connecticut, strict liability has also been imposed under an unfair trade practices statute.[20] However, strict liability was not imposed under either theory on a former Massachusetts property owner who rented an apartment to a childless man and wife and who later sold the property to a buyer who agreed to assume full responsibility for the property but did not inspect for or abate lead-based paint when requested to do so by the wife who had become pregnant.[21] In Maryland, strict liability under a consumer protection statute prohibiting landlords from engaging in unfair or deceptive trade practices was rejected in the absence of evidence that the landlord knew or had reason to know of the existence of lead-based paint in the leased premises.[22]

Breach of Lease Obligations. Lawsuits against landlords can be based on breach of a lease provision such as failure to make promised repairs. However, the typical lease requires tenants to accept the leased premises "as is" and requires the tenants to perform ordinary repairs, such as painting, and most tenants comply with the obligation. Landlords will sometimes agree to perform structural repairs, but will usually require tenants to give notice of a need for the repairs. In New York, however, there is a court decision that a lease requiring a tenant to keep the premises in good order and repair did not require a tenant to bear the cost of encapsulating or removing asbestos-containing material mandated by a supervening change of government policy.[23] That decision may be cited as precedent in disputes between landlords and tenants involving the cost of abatement of lead-based paint.

Breach of Warranty of Habitability. In New York and other states, the law may imply a warranty of habitability upon the sale of a new dwelling.[24] However, since lead-based

---

[19] Bencosme v. Kokoras, 400 Mass. 40, 507 N.E. 2d 49 (1989); Hardy v. Griffin, 41 Conn. Sup. 283, 569 A. 2d 49 (1989); See also Smith v. Coldwell Banker, 122 F. Supp. 2d 267 (D. Conn. 2000).

[20] Hardy v. Griffin, supra.

[21] Underwood v. Risman, 605 N.E. 2d 832 (Mass. 1993).

[22] Hayes v. Hambruch, 841 F. Supp. 706 (D. Md. 1994).

[23] Wolf v. 2539 Realty Associates, 161 A.D. 2d 11 (1st Dept., N.Y., 1990).

[24] De Roche v. Dame, 75 A.D. 2d 384 (3d Dept., 1980).

paint has not been sold for consumer use for many years, a valid claim for breach of such a warranty is unlikely. In New York, the law does not imply a warranty of habitability upon a sale of a used dwelling.[25]

In New York, lawsuits have been attempted based on a statutory warranty of habitability in section 235-b of the Real Property Law. The warranty, which cannot be waived or modified, imputes a covenant and warranty that the premises are fit for human habitation and the uses reasonably intended by the parties and that the occupants shall not be subjected to conditions dangerous, hazardous or detrimental to their life, health or safety. However, in a lawsuit involving lead-based paint, it was held that the statutory warranty is contractual in nature and did not create a new strict liability in tort excusing plaintiffs from proving the elements of negligence such as notice of a defect.[26] Subsequently, the state courts have rejected other tort claims seeking to recover for personal injuries or property damage.[27] Thus, the statutory claim must be for contractual damages equal to the difference between the fair market value of the premises, if they had been as warranted, as measured by the rent reserved under the lease, and the value of the premises during the period of the breach.[28]

Violation of a Statute, Ordinance or Code. The prior discussion under the caption "Strict Liability" includes strict liability based on failure to comply with statutory and regulatory requirements. Where a local ordinance imposed an affirmative duty to maintain premises in a safe condition, the New York Court of Appeals held that liability was not imposed without fault, but required a showing of common law negligence. The plaintiff must establish that the landlord had actual or constructive knowledge of both the hazardous lead condition and the residency of a child six years of age or younger. However, a statutory obligation to abate lead-based paint

---

[25] Young v. Keith, 112 A.D. 2d 625 (3d Dept. 1985). However, see Prugh v. Holmes, 405 A.2d 897 (Pa. 1979) for Pennsylvania Law.

[26] Mahlman v. Yelverton, 109 Misc. 2d 127. 439. N.Y.S. 2d 568 (Cir. Ct., Queens Cty., 1980).

[27] Curry v. Westchester Country Club, 186 A.D. 2d 712, 589 N.Y.S. 2d 491 (2d Dept., 1993); Carpenter v. Smith, 191 A.D. 2d 1036, 595 N.Y.S. 2d 710 (4th Dept., 1993); Stone v. Gordon, 211 A.D. 2d 881, 621 N.Y.S. 2d 220 (3d Dept., 1995).

[28] Park West Mgt. Corp. v. Mitchell, 47 N.Y. 2d 316, 391 N.E. 2d 1288 (1979); cert. den. 444 U.S. 992 (1979).

complied with authority enter a dwelling unit to inspect and abate, impose constructive notice on a landlord of any lead paint hazard in an apartment that the landlord knows is occupied by a child of the specified age.[29]

Nuisance (Private). Plaintiff attorneys frequently include claims of private nuisance in a complaint relating to lead-based paint. If properly pleaded, these claims can survive a motion to dismiss. However, the necessity to plead intentional and unreasonable conduct or negligence as well as other elements means that a plaintiff attorney is unlikely to succeed on a nuisance theory in any case which would not be successful on other theories.[30]

Damages. Plaintiff attorneys frequently seek both compensatory and punitive damages against landlords and other property owners. Where a jury found that a child was afflicted with "a rather severe neurological impairment," known as attention deficit disorder, which manifests itself in being hyperactive, impulsive and easily distracted, and had other learning and emotional disorders, a reduction of the jury's award from $1,700,000 to $200,000 by the trial court was reversed on appeal.[31] Punitive damages may also be awarded against landlords who commit willful violations or failure to control or abate lead-based paint after being issued a notice of violation.[32]

Some courts have issued a remittitur granting a new trial unless plaintiffs accepted a reduction of a jury verdict that the court deemed excessive.[33] One plaintiff refused a reduction and lost the entire case on appeal.[34]

---

[29] Juarez v. Wavecrest Management, 88 N.Y.2d 628, 649 N.Y.S.2d 115 (1996).

[30] Copart Industries v. Consolidated Edison, 41 N.Y. 2d 564, 362 N.E. 2d 968 (1977).

[31] Miller v. Beaugrand, 169 A.D. 2d 537, 564 N.Y.S. 2d 390 (1st Dept. 1991).

[32] Bencosme v. Kokoras, supra.

[33] Hiraldo v. Khan, 699 N.Y.S.2d 456 (2d Dept. 1999); Davis v. City of New York, 693 N.Y.S.2d (2d Dept. 1999).

[34] Arce v. New York City Housing Authority, 696 N.Y.S.2d 67 (2d Dept. 1999).

D. Investigation and Discovery

Defense attorneys seeking to investigate and discover evidence to evaluate the merits of claims made on behalf of a child face some initial challenges. Efforts to obtain information may be resisted as violations of privacy and other rights of the child and other family members. The resistance may come from government agencies, hospitals, medical professionals and others having custody of records as well as from counsel for the child or family members. Defense attorneys must be prepared to support requests for information.

Defense attorneys should assure that requests meet demonstrable standards of relevancy and reasonable scope. As shown in the studies and surveys listed in Chapter IV, Appendix A, there is ample scientific basis for discovery within a reasonable scope of medical, educational and other information relating to the plaintiff child and his or her family. Dr. David Bellinger, who participated in many epidemiological studies to establish an association between low level lead exposure and adverse health effects in children, has acknowledged that sociodemographic factors are much stronger predictors of child development than is low-level lead exposure.[35]

Both defense and plaintiff attorneys should be familiar with the discussion of the right of privacy and its limits by the U.S. Supreme Court in Whalen v. Roe.[36] In upholding a New York statute which required physicians to report prescriptions for certain drugs (including patient identities) to the State Department of Health, the Court noted that a physician-patient evidential privilege is unknown to the common law. In states where it exists by legislative enactment, it is subject to many exceptions and to waiver for many reasons. The Supreme Court discussed the right of privacy at length and held that it was not violated by the New York statute which contained detailed provisions designed to prevent misuse and improper disclosure.

Defense attorneys seeking information in federal government records should be familiar with the Freedom of Information Act (FOIA).[37] FOIA requires federal agencies to make

---

[35] "Low Level Lead Exposure and Cognitive Function in Children," Bellinger, Pediatric Annals, 23:11 (Nov. 1994).

[36] 429 U.S. 589 (1977).

[37] 5 U.S.C. § 552.

their records available to the public, subject to a number of exemptions which the courts construe narrowly. The exemptions include records or information compiled for law enforcement purposes to the extent that their production could reasonably be expected to constitute an unwarranted invasion of personal privacy. Plaintiff attorneys seeking to resist disclosure may also submit arguments based on the Privacy Act.[38] However, the Privacy Act allows access to documents if ordered by a court of competent jurisdiction.

Both defense and plaintiff attorneys should be familiar with state laws similar to FOIA and the Privacy Act that affect rights to records of state government agencies. For example, Connecticut has a Freedom of Information Act.[39] Connecticut has several laws that protect the privacy of medical, insurance, employment and other records. New York has a Freedom of Information Law.[40] New York also has a Personal Privacy Protection Law.[41] New Jersey has a Right-to-Know Law.[42] New Jersey has various statutory provisions regarding privacy of school and other records.

In a state court decision involving an alleged lead poisoned child, the Court of Appeals of Maryland reversed an order of the Baltimore City Court directing the Baltimore City Department of Social Services to disclose to a defendant landlord the entire files relating to the child and his parents. The Department cited statutory provisions declaring the records to be confidential and subject to release only pursuant to a court order. The Court of Appeals held that the defendant had made an adequate proffer demonstration showing need to inspect the records, but ruled that disclosure of the records should not exceed that need. The Court of Appeals remanded the case to the City Court for further proceedings that would accomplish the inspection

---

[38] 5 U.S.C. § 552a.

[39] 1 C.G.S.A., § 1-15 et seq.

[40] Public Officers Law, § 84 et seq.

[41] Public Officers Law, § 91 et seq.

[42] N.J.S.A. § 47:1A-1 et seq.

pursuant to a protective order which it suggested might be accomplished by an expanded in camera proceeding.[43]

An extensive discussion of investigation and discovery was presented in "Legal Hurdles to the Investigation and Discovery of Lead-Based Paint Cases," Carter and Leslie, Mealey's National Lead Litigation Conference, 1994.

There have been several recent decisions on discovery of medical records by the state courts in New York. In New York, the courts have discretionary authority on issues such as discovery of medical records.[44] Thus, medical records of family members of a plaintiff child have sometimes been denied.[45] Discovery of medical records of a plaintiff child's mother while pregnant with the child was allowed because the child's medical records are not separate from those of the mother during pregnancy.[46] One court authorized discovery of academic records of a mother and siblings and required the mother to take an IQ test.[47] However, other courts have refused to order a mother to take an IQ test.[48] Another court denied discovery of a mother's family history and high school transcripts.[49]

E. Experts and Their Qualifications and Testimony

Plaintiff attorneys customarily allege that a plaintiff child has suffered lead poisoning based on test results showing a blood lead level in excess of at least the 10 ug/dL lowest concern level discussed in the 1991 CDC Statement. They may also allege that the child has received medical attention including treatment by means ranging from monitoring to

---

43 Department of Social Services v. Stein, 328 Md. 1, 612 A.2d 880 (1992).

44 Andon v. 302-304 Mott Street Associates, 94 N.Y.2d 740, 709 N.Y.S.2d 873 (2000).

45 Mendez v. Equities by Marcy, 2005 N.Y. App. Div. LEXIS 13734 (1st Dept. Dec. 6, 2005); Alexander v. Westminster Presbyterian Church, 702 N.Y.S.2d 727 (4th Dept. 1999); and Monika W. v. Milovoi, 685 N.Y.S.2d 231 (1st Dept. 1999).

46 Lamy v. Pierre, 2006 N.Y. App. Div. LEXIS 9410 (2nd Dept., July 18, 2006).

47 Anderson v. Siegel, 680 N.Y.S.2d 587 (2d Dept. 1998).

48 Andon v. 302-304 Mott Street Associates, 257 A.D.2d 37 (2d Dept. 2000); Van Epps v. County of Albany, N.Y.L.J. p.1, April 26, 2000.

49 Helmer v. Draksic, No. 2003-4822 (Sup. Ct. Erie County, July 19, 2006).

hospitalization and chelation. They will usually also allege that the children are suffering neurological damage as evidenced by poor performance in school and on standard tests, behavioral problems, health problems, hyperactivity or attention deficit disorder.

Since the health problems alleged could have many causes other than lead exposure, plaintiff attorneys rely on witnesses having various qualifications as experts in various fields to prove causation. Defense attorneys rely on witnesses with various qualifications as experts to challenge the testimony of the plaintiff attorneys' witnesses and to rebut causation. The result at a trial will often rest on acceptance by the judge and jury of the testimony presented as expert by the plaintiff attorneys or the defense attorneys.

Witnesses claiming expert qualifications are not automatically allowed to testify in lawsuits involving lead exposure. They must meet the "Daubert" standards.[50]

Plaintiff experts begin with some significant challenges. While many epidemiological and other studies report an association between lead exposure and intelligence deficits or behavioral problems, the biological mechanisms by which lead could cause the injuries alleged have not been fully demonstrated although studies have suggested possible answers. Further, there is extensive medical literature showing that the health problems alleged to result from lead exposure can also result from a multitude of other causes unrelated to lead. The other causes include health problems inherited from parents; birth difficulties; illnesses and injuries sustained after birth; diet deficiencies; physical and mental abuse or stress encountered in family, neighborhood and school associations; exposure to medications, tobacco smoke, drugs and alcohol; food and other allergies; and lack of education as a result of nonattendance, negative attitude toward education or poor teaching. Before testifying, a plaintiff expert should carefully investigate all factors which could cause or contribute to a child's health or behavioral problems in order to be ready for cross-examination.

Both plaintiff experts and defense experts can be questioned on their professional qualifications, the scientific methods used in their field, their familiarity with the health history

---

[50] Campbell v. Metropolitan Property and Casualty Ins., 239 F.3d 179 (2d Cir. 2001); Daubert v. Merrell Dow, 509 U.S. 579, 125 L.Ed. 409 (1993); Frye v. U.S., 293 F. 1013 D.C. Cir. 1923).

of the plaintiff child, and the methods they used to determine that lead exposure did (or did not) cause or contribute to the child's alleged health or behavioral problems.[51]

To examine or cross-examine a purported expert, an attorney should know, among other things, the differences between a psychiatrist, a psychologist, a neurologist, a neuropsychologist, a toxicologist and other professionals working with children's health. The attorney should inquire about the witness's reliance on published epidemiological studies, health surveys and government reports and his or her familiarity with scope limitations, confounding factors and other matters affecting their validity or relevance. In one decision, a federal district court accepted expert testimony of two neuropsychologists testifying for the plaintiff on injury and injury causation rather than testimony of defense experts which included a medical doctor.[52]

The witness should be asked about the extent of his or her investigation of the child's health background and familiarity with studies that have shown that some children can deceive adults, including professionals, by playing dumb or staging a display of bad behavior. The witness should be asked about steps taken to verify that alleged health or behavioral problems are genuine. The witness should also be asked about steps taken to verify that health problems were caused by lead exposure rather than other known causes of the same problems. Expert testimony was rejected when the expert never examined the infant and was uncertain whether an initial high blood lead level was determined by a "finger stick" test, especially when a later test showed a level of only 3 ug/dL.[53]

NOTE: As this book went to press, an appellate court in New York held that reports by a psychologist and a speech pathologist were admissible in evidence as business records. Wilbur v. Lacerda, 2006 N.Y. App. Div. LEXIS 14291 (2d Dept. 2006).

---

[51] Campbell v. Metropolitan Property Casualty Ins., supra.

[52] Dombrowksi v. Gould Electronics, 85 F. Supp. 456 (M.D. Pa., 2000). See also Walton v. Albany Community Dev. Agency, 718 N.Y.S.2d 456 (3d Dept. 2001).

[53] Arce v. New York City Housing Auth., 696 N.Y.S.2d 67 (2d Dept. 1999).

F. Lawsuits Against Liability Insurers

Many lawsuits are being commenced against liability insurers seeking defense and indemnification against liability for bodily injury claims for exposure of children to lead-based paint. Lawsuits are also being commenced against liability insurers seeking defense and indemnification against liability for alleged property damages resulting from a requirement to abate lead-based paint. Lawsuits are also being commenced against casualty insurers to recover for losses resulting from a need to abate lead-based paint.

When a property owner seeks defense and indemnification from an insurer under a liability policy for a child's lead poisoning claim, a threshold question is whether the child's bodily injury occurred during the policy term. If a child was discovered to be afflicted by lead poisoning prior to the policy term, the courts may decline to enforce coverage even though the child's exposure to lead-based paint and resulting injury continued into the policy term.[54] However, evidence of new injuries due to continued exposures during the policy period may trigger coverage.[55]

If a complaint alleges claims apparently within policy coverage, an insurer must defend its policyholder unless and until it is determined that the policy does not obligate the insurer to indemnify the policyholder against liability. If the insurer wishes to clarify its obligations, it can seek a declaratory judgment.[56] In other words, the duty to defend arises when a claim is presented initially, not after the validity of the claim has been established.[57] The obligation to defend is also not excused by allegations in a complaint against a former

---

[54] Harford Mutual Insurance Company v. Jacobson, 73 Md. App. 670, 536 A.2d 120 (1988); St. Leger v. American Fire, 870 F. Supp. 641 (E.D. Pa. 1994), affd 61 F.3d 986 (3d Cir. 1995).; Hartford v. Estate of Turks, 206 F. Supp. 2d 968 (2002).

[55] Scottsdale Insurance v. American Surplus, 81 F. Supp. 210 (D. Md. 1993).

[56] Harford Mutual Insurance Company v. Jacobson, supra. St. Leger v. American Fire, supra; Oates v. State of New York, 597 N.Y.S.2d 550 (Ct. Cl. 1993).

[57] NL Industries, Inc. v. Commercial Union, 828 F. Supp. 1154 (D.N.J. 1993), reversed on choice of law issue, 65 F.3d 314 (3d Cir. 1995). In subsequent proceedings, the federal district court applied New York law, rather than New Jersey law, and again found that the defendant insurer was obligated for defense costs. 926 F. Supp 446 (D.N.J.1996).

manufacturer of lead-based pigments that include claims that the manufacturer's conduct was intentional.[58]

Many general liability insurance policies contain a pollution exclusion in which pollutants are defined as "any solid, liquid, gaseous, or thermal irritant or contaminant, including smoke, vapor, soot, fumes, acids, alkalis, chemicals and waste." Some courts have ruled that lead-based paint is not a pollutant within this definition or that the exclusion is ambiguous, thus requiring an insurer to provide coverage consisting at least of a duty to defend.[59] However, other courts have held that the pollution exclusion applies to lead-based paint claims.[60]

In U.S. Liability v. Bourbeau,[61] a lawsuit to recover damages for soil contamination caused by improper abatement of lead-based paint, the First Circuit offered the following explanation of the apparent conflict in the court decisions interpreting of the pollution exclusion:

> Most importantly, McFadden was not an environmental pollution case. McFadden concerned personal injury caused by the presence of lead paint in a household. This case concerns injury to property caused by the alleged negligent discharge of lead paint onto property. The latter is a classic example of "pollution"--the discharging of a harmful substance onto land--while the former is most demonstrably not. An objectively reasonable person simply would not ascribe the word "pollution" to the presence of paint in a house.

---

[58] NL Industries, Inc. v. Commercial Union, supra.

[59] Lefrak Org. v. Chubb Custom Ins. Co., 942 F. Supp. 949 (1996); Atlantic Mutual v. McFadden, 413 Mass. 90, 596 N.E. 2d 762 (1992); Generali-U.S. Branch v. Caribe Realty, 612 N.Y.S. 2d 299 (St. Ct., N.Y. County, 1994). See also Schumann v. State of New York, 610 N.Y.S. 2d 987 (Ct. Cl. 1994).

[60] Peace v. Northwestern Nat'l Ins. Co., 228 Wis.2d 106, 596 N.W.2d 429 (1999); Oates v. State of New York, 610 N.Y.S. 2d 987 Ct. Cl. 1993); St. Leger v. American Fire, supra; Kaytes v. Imperial Casualty, No. 93-1573 (E.D. Pa. Jan. 6, 1994); U.S. Liability v. Bourbeau, 49 F.3d 786 (1st Cir. 1995).

[61] U.S. Liability v. Bourbeau, 49 F.3d 786 (1st Cir. 1995).

When a primary liability insurance policy contained an exclusion for lead-based paint, an insurer which issued a "follow the form" umbrella excess policy was not required to drop down and provide coverage.[62]

Over nine months delay of notice to an issuer justified refusal of coverage. Notice given to a broker was not notice to an insurer.[63]

In 2005, a federal district court in New Jersey denied a liability insurer's motion for summary judgment that it was not required to defend or indemnify a painting contractor in a lawsuit by homeowners who alleged that they suffered bodily injury and property damage resulting from exposure to lead, toxic fumes and dust. The insurer based the motion, among other things, on a total pollution exclusion from the bodily injury coverage and the property damage coverage and on a lead paint or lead contamination exclusion from the bodily injury coverage. The court held that the total pollution exclusion did not apply to a claim for the cost of removal and replacement of a top layer of soil contaminated with paint chips. The court held that the lead paint or lead contamination exclusion applied to bodily injury, but not property damage.[64]

In 2005, a trial court held that a liability insurer was obligated to defend and indemnify a property owner and a property management company although the policy language limited coverage to a blood lead concentration of 15 mg/dL. The court held that the policy language was ambiguous because the coverage limitation did not appear in the insuring language or the exclusions, but in a definition of documentation. The court concluded that the insured could not reasonably be expected to look for an important exclusion in a definition of documentation.[65]

G. Lawsuit Against a Hospital and a Doctor

A trial court in New York recently refused to grant defense motions for summary judgment in a lawsuit against the New York City Health and Hospitals Corporation and a doctor

---

[62] Westsview Associates v. City of New York, 693 N.Y.S.2d 67 (1st Dept. 1999).

[63] Sphere Drake Inc. v. YI Realty, 1999 U.S. Dist. LEXIS 13419 (Aug. 31, 1999).

[64] Merchants Ins. Co. v. Hessler, 2005 U.S. Dist. LEXIS 18173 (D.N.J., 2005).

[65] United States Fidelity & Guaranty v. Cruz, 8 Misc. 3d 1015, 801 N.Y.S. 2d 792 (2005).

for alleged failure to advise the child's mother how to avoid lead exposure and failure to administer chelation treatment because there were material unresolved fact issues.[66]

H. Lawsuit Against a Corporate Officer

In a recent case in New York, claims were made against a corporate landlord and an officer of the corporation that a plaintiff child had been injured by lead-based paint. A motion for summary judgment by the corporate officer was denied because of an issue of fact whether the officer had made a misrepresentation about use of lead-based paint in the apartment to the child's mother.[67]

I. Lawsuits Under the Lead-Based Paint Poisoning Prevention Act and the Residential Lead-Based Paint Hazard Reduction Act

Summaries of court decisions in lawsuits under the Lead-Based Paint Poisoning Prevention Act and the Residential Lead-Based Paint Reduction Act are provided in Chapter XVI. They include recent decisions rejecting efforts to claim private rights of action under these laws against federal, state and local government and their housing instrumentalities.

J. A Recent Article

An article was recently published in the Toxics Law Reporter titled "Getting the Lead Out? The Misuse of Public Nuisance Litigation by Public Authorities and Private Counsel."[68] Written from the point of view of defense counsel, the article provides information about the history of lead and its regulation and about lead litigation. The article criticizes the recent public nuisance lawsuits by state attorneys general, especially the jury instructions given at the second trial in Rhode Island v. Atlantic Richfield Co.

---

[66] S.S. v. New York City Health and Hospitals Corporation, Sup. Ct., N.Y. County, New York Law Journal, April 10, 2006, p. 18, col. 1.

[67] Espinosa v. Rand, 2005 N.Y. App. Div. LEXIS 13626 (1st Dept., 2005).

[68] Toxics Law Reporter, Vol. 21, pp. 1071-1098, 1124-1152, and 1172-1196 (2006).

ACKNOWLEDGEMENTS

The author wishes to thank James F. Johnson, M.D., FAAP, who contributed valuable research materials and comments on the health effects of lead. Dr. Johnson, who has many years of experience as a pediatrician, recommends that the scientific controversies described in Chapter IV be viewed in light of the primary goal to protect children.

The author also wishes to thank Daniela Capraru who performed many weeks of paralegal work including research, proofreading, correcting, organizing and revising the text, and asking many good questions that led to clarifications and improvements.

The author also thanks Marilyn Gibbs who is an assistant and secretary at Windels Marx Lane & Mittendorf and also helped with the first edition of the Lead Regulation Handbook. Her help in typing the scientific and legal text, organizing the numerous footnotes, and other assistance is much appreciated.

Edward E. Shea

INDEX